71+10 N

Science Projects

Dr C. L. Garg

Senior Scientist, DRDO

Ministry of Defence

Dr Amit Garg

PhD (Electronics)

PUSTAK MAHAL ®

Delhi • Bangalore • Mumbai • Patna • Hyderabad

Publishers

Pustak Mahal®, Delhi-110006

Sales Centres

- 6686, Khari Baoli, Delhi-110006, *Ph:* 23944314, 23911979
- 10-B, Netaji Subhash Marg, Daryaganj, New Delhi-110002
  Ph: 23268292, 23268293, 23279900 • *Fax:* 011-23280567
  E-mail: rapidexdelhi@indiatimes.com

Administrative Office

J-3/16 (Opp. Happy School), Daryaganj, New Delhi-110002
Ph: 23276539, 23272783, 23272784 • *Fax:* 011-23260518
E-mail: info@pustakmahal.com • *Website:* www.pustakmahal.com

Branch Offices

BANGALORE : 22/2, Mission Road (Shama Rao's Compound),
Bangalore-560027, *Ph:* 22234025 • *Fax:* 080-22240209
E-mail: pmblr@sancharnet.in • pustak@sancharnet.in

MUMBAI : 23-25, Zaoba Wadi (Opp. VIP Showroom), Thakurdwar,
Mumbai-400002, *Ph:* 22010941 • *Fax:* 022-22053387
E-mail: rapidex@bom5.vsnl.net.in

PATNA : Khemka House, 1st Floor (Opp. Women's Hospital), Ashok Rajpath,
Patna-800004, *Ph:* 3094193 • *Telefax:* 0612-2302719
E-mail: rapidexptn@rediffmail.com

HYDERABAD : 5-1-707/1, Brij Bhawan, Bank Street, Koti,
Hyderabad-500095, *Telefax:* 040-24737290
E-mail: pustakmahalhyd@yahoo.co.in

ISBN 81-223-0150-9

Illustrated by : **Manish (Innovator)**

14th Revised and Enlarged Edition : August 2003
17th Reprint Edition : April 2005

Printed at: Unique Colour Carton, Mayapuri, Delhi-110064

Preface _____

Scientific projects and models are an integral part of science education today. To harness the skills of students, models and projects play a vital role and carry much weight in their overall performance, particularly in the 9th, 10th, 11th and 12th standards. Preparing working models has rightly been made compulsory in schools for all major branches of science: Physics, Chemistry, Biology, Botany and Electronics. Working on these models helps students better grasp the basic principles of science involved in the functioning of each model.

Considering these facts, we have painstakingly compiled a series of ideas for preparing models/projects that could be adapted to suit each student's syllabus. These projects were successfully tested beforehand. Besides, all components used are Indian and easily available in the market.

But in today's highly competitive times, a book is simply not enough. Therefore, for the first time, Pustak Mahal presents a Rapidex self-learning kit comprising the book **71 + 10 New Science Projects** and an audio-visual CD. In an easy-to-use format, the kit has been conceived and designed to meet students' and other users' aspirations. The graphically animated objects and methods facilitate quick learning.

In addition to meeting the needs of students, this Rapidex self-learning kit is an ideal choice for hobbyists and parents seeking to inculcate a scientific temperament in their children. The projects dealt in this Rapidex self-learning kit are unique but easy to handle.

Furthermore, the scientific explanations in these projects make understanding easier and increase IQ. The 10 new projects in the 15th edition make the book all the more informative and interesting.

My sincere gratitude to all the books that were source material and to my son Rajeev Garg for his help in preparing the manuscript. This revolutionary Rapidex self-learning kit will serve as an indispensable tool for students, hobbyists and other amateur scientists.

— **Dr C.L. Garg**

Preface to the 14ᵗʰ Edition ————————————

The book entitled '71 Science Projects' has been revised by replacing an old project and adding 10 new electronics projects. These electronics projects will be quite useful for 10 + 2 science students. Now the book has been renamed as '71 + 10 New Science Projects'.

71 + 10 New Science Projects
About the Tutorial CD

This computer-based tutorial is a unique feature of the book. Brought to India for the first time, it helps you understand project-making more easily. This CD contains 50 projects covering all the subjects. Fairly simple projects or variations of projects already included (and with the same props) have been left out. All projects requiring concept building and proper understanding of facts are covered. This CD is a bonus that provides true value for money.

Projects _____

1. Making and controlling a diver

A diver is a person who explores the underwater world by diving. Divers explore the oceans, lakes and rivers by taking deep dives under the water. They make use of diving suits, breathing tubes, etc. Divers do many important jobs such as studying plant and animal life at the bed of the sea extracting minerals, or even saving people from drowning.

You Require
- *A glass bottle with a wide mouth*
- *A tight-fitting cork*
- *A flat piece of plastic*
- *A plastic tube*
- *Adhesive*
- *Thin wire*
- *Scissors*
- *Modelling clay (plasticin)*

What To Do
- Take an empty glass bottle with a tight-fitting cork. If the cork is dry and rigid, leave it in the water until it becomes flexible, so that it can be pushed down into the mouth of the bottle.

- Take a flat plastic sheet and draw the outline of the diver, as shown in the figure. The sheet should be thin enough to fit into the glass bottle. Use scissors to cut the diver to shape.

- Take a small piece of the plastic tube from an old ball-point pen. Seal one of its ends with modelling clay or plasticin. The other end should remain open. Glue this tube to the diver with the adhesive (Quickfix) as shown in the diagram. The air in the tube will make the diver float.

- Wind some wire around the diver's feet. The weight of this wire will make it stay upright in the water. Use enough wire to make the diver sink. Then remove a few turns of the wire so that the diver just floats at the bottom.

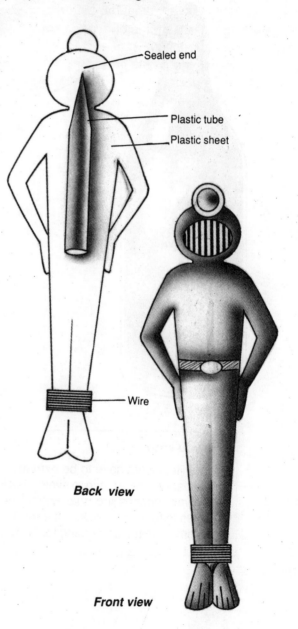

Sealed end

Plastic tube

Plastic sheet

Wire

Back view

Front view

7

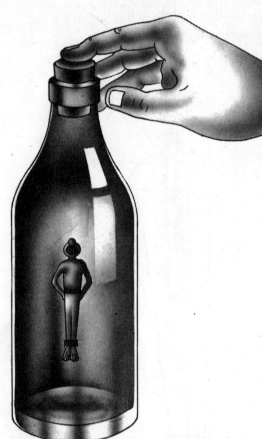

- Put the diver in the glass bottle and fill the bottle with water.

- Push the cork into the bottle. This increases the pressure in the water and some water finds its way into the plastic tube. Consequently, the diver sinks.

- Pull the cork a little out of the mouth to make the diver rise. With a careful adjustment of the cork, the diver can be made to stop at any depth you want.

- By the pressure command of your finger, this model diver will move up or down in the water. You can even make the diver hover at any depth.

Did you know...

In real life, divers have to be extremely careful about the speed at which they come back to the surface to avoid what is called "*the bends* or *caisson disease*". If a diver surfaces too quickly, the reduced water pressure causes nitrogen bubbles to form in his blood-stream. This results in horrible pain, paralysis and sometimes death. By coming to the surface slowly, divers avoid this condition.

2. Making an abacus

The abacus is a device once widely used for counting. An abacus has beads that are moved left and right on strings that are tied in a frame. The beads on the bottom string represent the value of units. Those on the second string have the value of tens. The beads on the third string represent hundreds and so on. One can master the abacus by different movements of the beads on the strings. Once the abacus is mastered, a person can add, subtract, multiply and divide quickly by moving the beads on the strings.

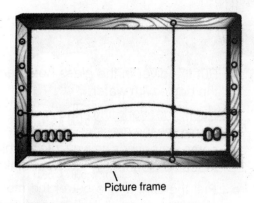

Picture frame

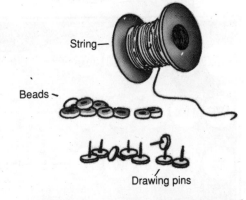

Drawing pins

You Require
- *Picture frame*
- *String*
- *Thumb tacks or drawing pins*
- *Beads or buttons*

What To Do
- Take a picture frame measuring 30 cm long and 22 cm wide.

- Cut 5 pieces of string long enough to cross the picture frame with an extra 7.5 cm on each end for making knots.

- Fix 5 evenly spaced drawing pins along each side of the frame.

- Slide 7 beads or buttons on each string.

- Tie the end of one string to the first pin at either side of the frame. Continue tying the strings until you have tied all the five across the frame.

- On all strings, move 5 beads to the left side of the frame and 2 beads to the right side.

- Now take an additional piece of string more than twice the width of the frame to make the dividing string of the abacus. Tie this piece around the frame at the top about halfway across, as shown in the figure.

- With the dividing string tied to the top of the frame, bring it down to the next string and tie a knot. Continue down to the next string and tie another knot. Work down to the bottom of the frame until knots have been tied with all the strings. Then tie the string around the bottom of the frame. You now have the dividing line (string) of the abacus.

Your abacus should now have 5 strings extending from left to right, with five beads to the left side of the dividing string and 2 beads to the right as shown in the Fig. A of project 2B.

■■

3. Using your abacus for calculations

Hold the abacus before you. The bottom string represents units. The five beads on the left have a value of 1 each. The two beads on the right have a value of 5 each.

The second string represents values of tens. Each bead on the left represents 10 units. Each bead on the right represents 50 units.

The next string represents hundreds, the fourth string thousands and fifth string 10,000s, as shown in Figure A.

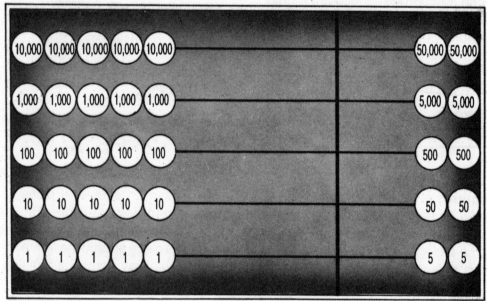

Fig. A

How to use the abacus for calculations
The abacus constructed in this way is now ready for use. You can try the following calculations.

- Count 4 on your abacus: To count 4, push 4 beads from the left side of the bottom string to the Centre (Fig. B).

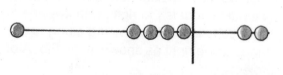

Fig. B

- Now add 4 and 5. To do this, push 4 beads from the left side of the bottom string to the centre (Fig. B). To add 5, push one bead from the right side to the centre (Fig. C). So you have: 4 + 5 = 9.

Fig. C

- Now subtract 6. Do this by moving the 5-unit bead on the bottom right back to the frame edge and move one unit bead on the left back to the frame edge (Fig. D). So you have: 9 — 6 = 3.

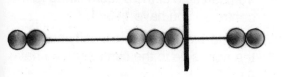

Fig. D

- Now add 8. To do this, move the 5-unit bead on the right to the centre and move three 1-unit beads from the left to the centre. You will find only two to move. Move them. Your abacus will look as shown in Fig. E.

Fig. E

- You have a value of 10. Move the five 1-unit beads on the left to the frame and move the one 5-unit bead on the right back to the frame. Replace their value of 10 by moving one 10-unit bead on the second string from the left to the centre. You still have a value of 10. Now move the one 1-unit bead to the centre (Fig. F). You have a total of 11 (3 + 8 = 11).

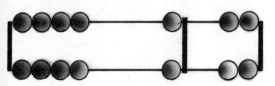

Fig. F

- Now add 295. For doing this, add 5 first by moving a 5-unit bead from the right to

Fig. G

the centre (Fig. G). Next, add 90 by moving a 50-unit bead from the second string right and four 10-unit beads from the left (Fig. H).

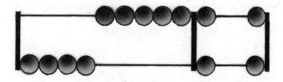

Fig. H

- Last, add the 200 by moving two 100-unit beads on the third string. You now have a total of 306 as shown in Fig. I.

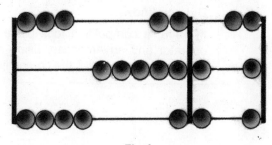

Fig. I

- Since the value on the second string is equal to 100, the abacus can be adjusted by moving the beads to the frame edges and replacing their value by moving one

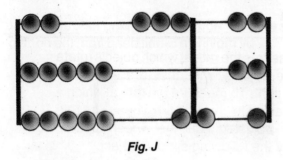

Fig. J

100-unit bead from the third string (Fig. J).

You can try some more calculations. After practising several times, you will be able to use the abacus very quickly and accurately.

Try the following with your abacus

(i)   Add 1 and 3 i.e. 1+3=4

(ii) Now add 5 to it i.e. 4+5=9

(iii) Now add 2 to it i.e. 9+2=11

(iv) Add 256 to it i.e. 11+256=267

(v)   Now subtract 50 from it i.e. 267–50=217

Solution

(i)   Move one bead and then three beads from left side of the bottom string to the centre so you have 1+3=4.

(ii) To add 5 to it, push one 5 unit bead from the right side to the centre so you have 4+5=9.

(iii) To add 2 to it, move one more 5 unit bead from the right side to the centre and then three single beads back to the frame edge, so you have 9+5–3=11.

(iv) To add 256, move both the 5 beads back and move one 10 bead to the centre so that again you have 9 total of 11. Now move two 100 beads, one 50 bead, one 5 bead and one unit bead to the centre so you have 11+256=267.

(v)   To subtract 50, you move back the bead of 50 so you have 267–50=217.

Did you know...

In 1946, a competition took place between a Japanese clerk, an abacus' operator and an American, using an electronic calculator. And surprisingly enough Japanese won the competition.

4. Making a stroboscope

A stroboscope is an instrument that measures the rate at which objects rotate or vibrate. It is also used to study the objects while they are moving. A stroboscope gives out bright flashes of light. To measure the rate at which a wheel is rotating, the stroboscope is directed to shine on the wheel. The rate of the stroboscope's flashing is then adjusted until the wheel appears stationary. This effect can be shown by the following simple project.

You Require
- *One cardboard disc 4" in diameter*
- *One hand drill*
- *One nail*

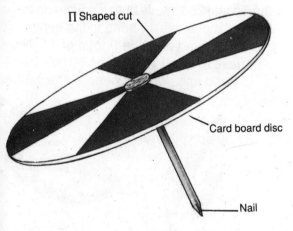

Π Shaped cut

Card board disc

Nail

What To Do
- Take a cardboard disc 4" in diameter Divide the circle into eight equal pie-shaped sections. This can be done by drawing a line through the centre of the circle and then drawing another line at right angle through it. This divides the circle into four equal parts. Now, divide each of these four parts into halves thus, producing eight sections. Leave one section white and colour the next. In this way, you will have four white sections and four coloured sections.

- Now, force a nail through the centre of the circular disc in such a way that the cardboard cannot turn unless you turn the nail. If necessary, you can glue the nail.

- Insert the nail into the chuck of the hand drill. Now, turn the handle of the drill. As you change the speed of the drill, the disc will appear as if it is standing perfectly still. If you reduce the speed slightly at this position, the disc will appear to be turning backwards. If you increase the speed, the section will move rapidly forward. This demonstrates the stroboscopic effect.

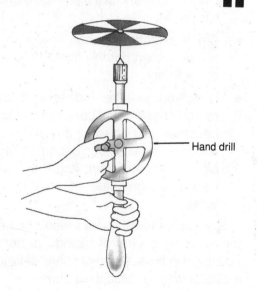

Hand drill

5. Making a weather-indicating flower

The amount of moisture in the air, i.e. humidity, can be used as the basis for forecasting the weather. When the humidity is high, rain is likely, while a low humidity usually indicates a dry spell. This project is based on the colour change produced in cobalt chloride by the moisture in air.

You Require

- *Cobalt chloride, about 20 gm*
- *Blotting paper*
- *Thread*
- *One thin wooden stick about 3 mm thick and 15 cm long*
- *Scissors*

What To Do

- Take a small sheet of blotting paper and cut it in the shape of a flower, as shown in figure.

- With the help of cotton thread and a sewing needle, tie it with the thin wooden stick.

- Now take about 20 gm of cobalt chloride from a chemicals supplier or from your laboratory and make a strong solution of it in water.

- Dip the blotting paper flower in this solution, then put the flower in a warm place to dry.

- Now, your weather-indicating flower is ready for use. If the colour of the flower becomes pink on some day, it indicates a storm; if the colour becomes bluish-pink, it indicates rain; and if the colour becomes light blue, it indicates dry weather.

Note: You can also make a weather indicator by using a strip of blotting paper and soaking it in the solution of cobalt chloride in a similar way as described above.

Flower made from a pink coloured blotting paper

Weather indicating flower

6. Demonstrating how clouds are formed

In the atmosphere, clouds form when air containing water vapour cools. The cooler air cannot hold so much water vapour, so some of the vapour condenses to form clouds. This project shows how to simulate this atmospheric phenomenon by forming a cloud in a glass jar.

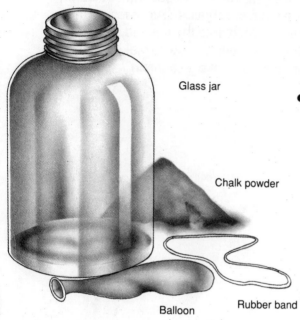

Glass jar

Chalk powder

Balloon

Rubber band

- Remove the lid and sprinkle some chalk powder or talcum powder inside the jar. Immediately, cover the jar with a thin rubber sheet cut from a large balloon and put a rubber band around the neck of the jar.

- Now, compress the air in the jar by pressing the rubber with your fist. The air becomes warmer and absorbs more water vapour. After 15 seconds, release the rubber. The air cools. Some of the water vapour condenses on the chalk dust, forming a miniature cloud of water droplets.

You Require
- *One large glass jar*
- *Fine chalk powder*
- *Thin rubber sheet cut from a balloon*
- *Rubber band*

What To Do
- Take a large clean glass jar and pour some water into it. Cover the jar with a lid and leave it for about 15 minutes. Some of the water will evaporate to form an invisible vapour.

7. Making a siphon fountain

A siphon is a bent tube with arms of unequal lengths. It is a useful device for transferring liquid from one container to another at a lower level. A rubber or plastic tube is normally used as a siphon. The end of the tube with shorter arm is placed in a container and the liquid is sucked up in to the tube. A finger is put on the open end of the tube. Keeping one end of the tube in the liquid, the open end is then placed in the second container, which should be at a lower level than the first container. Air pressure then forces the liquid to flow from the first container into the second container. This is how a siphon works. This project teaches us how to make a siphon work and how to use the siphon principle to make a fountain

You Require
- One clear plastic bottle
- One cork or rubber stopper
- Two plastic or glass tubes
- Two mugs
- Plasticin (toy's clay)

What To Do
- Take a clear plastic bottle about 8 to 10 inches in height and a cork or rubber stopper which fits into it tightly. Drill two holes through the cork into which plastic or glass tubes can be fitted tightly. One tube should project through the cork into the bottle by one inch, the other by six inches. If necessary, use glue to ensure an air-tight seal around the tubes, and seal the ends of the tubes with plasticin.

- Take two plastic mugs, one empty and the other half full of water. The mug containing water should be put on a stool about 6" to 12" above the empty mug.

- Carefully turn the bottle upside down so that the tube that projects the farthest into the bottle is over the mug that is half full of water.

- Remove the seal from the tube over the half-filled mug and immediately lower

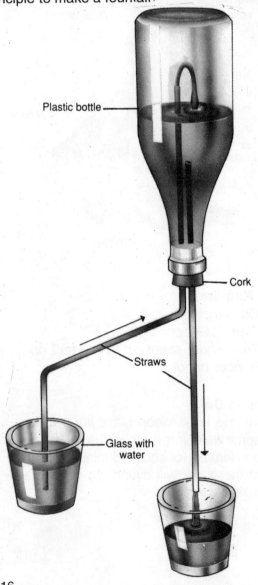

Plastic bottle

Cork

Straws

Glass with water

the end of the tube under the surface of the water. The water level in the mug should rise, while that in the bottle falls. After a few seconds, this flow of water should stop.

- If the tips of both the tubes are now showing above the water surface in the bottle, you either have a leak or are using a bottle that is too small. In either case, you will need to start again. If neither tube shows above the water surface in the bottle, you will need to introduce a little air into the bottle by lifting the tube out of the mug of water immediately.

- Now, remove the seal from the tube over the empty mug. Water will flow quite rapidly from the full mug up into the bottle, forming a fountain and then down into the empty jar.

Note: You can make a simple siphon as follows:

- Take a bucket of water and place it on a table. Take another empty bucket and place it on the floor.

- Place one end of a length of rubber tubing in the top bucket, and suck at the other end until the tube is full of water.

- Leave one end of the tube in the water and place your finger over the other end. Put this end of the tube into the empty bucket on the floor. On removing the finger from the tube, water will start to flow into the lower bucket. It will continue to flow as long as the top end of the tube is under water. This illustrates the basic principle of siphon.

■ ■

8. Making a model of an elevator

People are usually lifted to the tops of tall buildings by elevators or lifts. An elevator car is raised by a windlass that is turned by an electric motor. Tracks keep the car moving in a straight line. A weight balances the weight of the car so that it is easier to pull up the car. The motor of the elevator can turn the windlass either way. It can wind up the cable and raise the car. It can unwind the cable and let the car go down. Electric wires go from the car to the motor. A person in the car can make it go up, go down or stop by using switches in the car. You can make a model of an elevator as follows.

You Require
- *One plyboard sheet about 24" × 12"*
- *Six cotton thread reels*
- *Six nails*
- *One cardboard box*
- *String*
- *A small weight*
- *One toy boy and one toy girl*

Reel

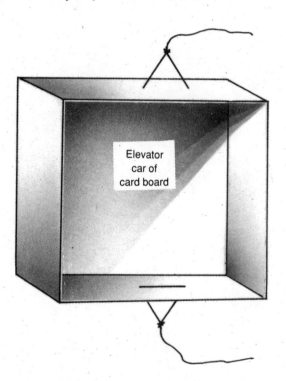

Elevator car of card board

What To Do
- Take a cardboard box, say a shoe-box, without the lid. Tie one string to the bottom of the box and two strings to the top of the box. This box will act as the car of the elevator.

- Drive nails through spools into the wooden board as shown in the figure.

- Slip the bottom string of the car over the spools marked 4, 3, 2 and 1 in the figure. Wind the string several times around spool 2. Tie its other end at the top of the box.

- Run the second string from the top of the box over spools 5 and 6 and finally fasten it to a weight of about 100gm to balance the weight of the car.

- Turn the windlass (spool 2) with your fingers. The car will go up and down. You can put two toys, one of a boy and the other of a girl, in the box as passengers.

You can make this model more attractive by putting it in a model building constructed of cardboard.

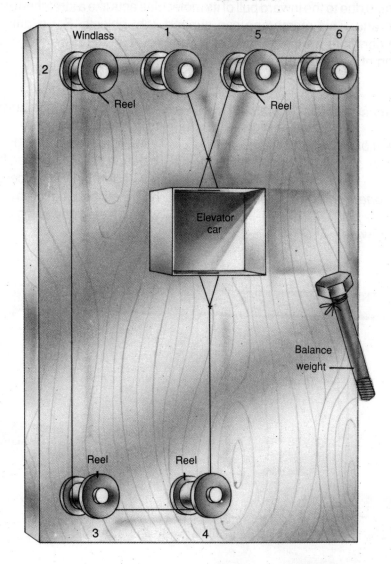

Note: The fastest lift of the world is fiffted in "sunshine 60" building of Japan. Its speed is 36.56 km/h.

9. Determining the surface tension of water

Surface tension is a force that occurs at the surface of a liquid. The surface of a liquid due to the inward pull of its molecules acts like a stretched elastic membrane. This membrane can support light objects. For example, a blade can float on the surface of water. You can measure the surface tension of water by this simple project.

You Require

- One thin plastic sheet 5cm × 5cm × 3mm
- Thread
- One wooden block 10cm × 10cm × 2cm
- Two wooden strips, .25cm × 1.5cm × 1cm each
- Iron nails, one iron ring of about 5cm diameter, one pin
- One bowl of water

What To Do

- To measure surface tension, you have to make a balance device like the one shown in the figure. For making the balance, take a wooden block and glue a wooden strip at its centre. Make the beam of the balance by using another wooden strip and using a pin as a pivot.

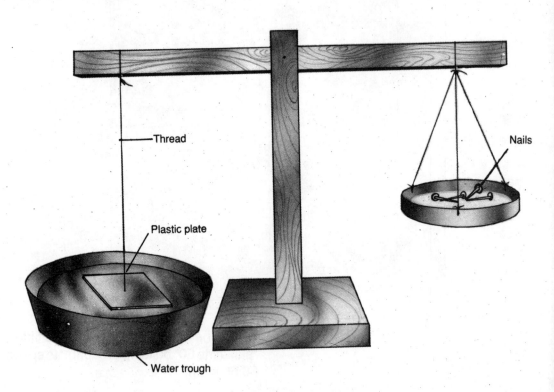

Thread

Plastic plate

Water trough

Nails

Balance made from wooden strips

- Take the iron ring and make a small bag of cloth. Tie three threads to make a pan of the balance as shown in the figure.

- Take the plastic sheet and put a tiny hole in the centre with a needle. Tie a knot in a thread and use it to suspend the plastic sheet from one end of the balance as shown in the figure.

- Use small nails on the other end to balance the plastic sheet.

- Now, bring the bowl full of water and put it under the plastic sheet in such a way that it just touches the surface of water. You will see that the balance is disturbed.

- Now, add small nails in the pan until the plastic sheet suddenly breaks away from the surface of water. This extra weight is nothing but a measure of the surface tension of the water — the force with which it attracts the plastic.

- You can repeat the experiment by using mustard oil, kerosene oil, alcohol, etc. instead of water and measure their respective surface tensions also.

Some examples of surface tension

- A razor blade or a sewing needle can be made to float on the surface of water, although these things are heavier than water. Put the blade or needle on a piece of blotting paper and then place it gently on the surface of water in a glass. The blotting paper soon gets drowned into the water and the blade or needle remains floating on the surface of water. This is due to surface tension.

- The liquid drops assume spherical shape due to surface tension.

- Some insects can run on the surface of water without wetting their feet. It is due to surface tension of water.

- A blotting paper absorbs ink due to capillarity which is also a phenomenon related to surface tension.

- The nib of a pen is slit at the tip to provide the continuous flow of ink by surface tension.

- Soil retains moisture due to capillarity.

■■

10. Making a spinning snowman

It is a science toy which works on the phenomenon of the surface tension of water. It makes use of mothballs, which weaken the pull of surface tension in the water close to them. The stronger pull of the surface tension in front of each mothball pulls the sticks and the snowman spins in a circle.

You Require
- *A plastic tub for water*
- *One big cork*
- *Four rectangular slabs of light wood (1" × 1" × 0.25")*
- *Four wooden sticks about 8 to 10 mm thick*
- *Four mothballs*
- *Paper*
- *Scissors*
- *Glue or tape*

What To Do
- Draw a small snowman on the paper, as shown in the figure, colour it and cut it out.

- Cut a slice of the cork about 1" thick and take four rectangular pieces of light wood. Make a small notch in each piece of wood. Glue one mothball in each notch in such a way that their surfaces remain in touch with water.

- Attach four sticks to the cork such that they are in the shape of a cross, as shown in the figure. At the free ends of these sticks, attach one wooden piece with glue, as shown in the figure.

- Stick the snowman to the slice of the cork.

- Take a plastic tub and fill it with water. Now, put the snowman into the tub of water and watch it. The whole assembly will spin around due to the surface tension of water. ■■

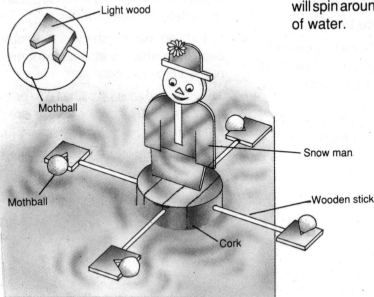

22

11. Making a hovercraft

A hovercraft is a vehicle which can run both on water and land with equal ease. It is pushed along by means of a propeller and steered like an airplane with a rudder. Actually, a hovercraft neither touches land nor water. It moves on a cushion of air. The air cushion is produced by a large flat fan which blows air downwards so hard that it lifts the craft off the ground or water. This cushion of air supports the craft so that there is no contact between it and the surface. There are no wheels and the craft can travel over land or water. In this project, the making of two simple models of hovercraft is described.

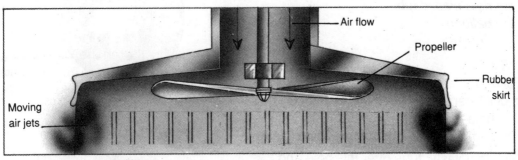

Hovercraft

You Require

- *A rectangular wooden board of the size 10cm × 8cm × 0.5cm*
- *One metal tube about 3" long with a hole of quarter inch*
- *Rubber tubing*
- *Hair drier with a rubber cap having a hole*

What To Do

- Take a wooden board measuring 10cm × 8cm × 0.5cm and drill a hole about one millimetre in diameter at its centre.
- Fix a metal tube (about 3" long with a 6mm diameter hole) over this hole with glue.

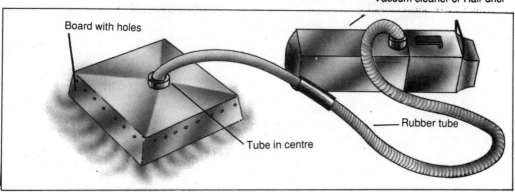

- Now, connect the metal tube to the rubber cap of the hair drier by a rubber or plastic tubing.
- Switch on the hair drier. As the air pushes through the hole in the board, it is slightly lifted up. Now, with your finger, give a little push to the wooden board. It will move.

An alternative simple model

You Require
- *One metal or plastic disc about 10 cm in diameter and 2 mm thick*
- *One rigid plastic tube*
- *One balloon*
- *One needle*

What To Do
- Take a metal or plastic disc about 10 cm in diameter and 2 mm thick. Drill a very fine hole (half mm in diameter) in the centre of it with the help of a needle.
- Glue the plastic tube over this hole and let it dry.

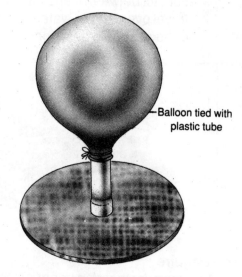

Balloon tied with plastic tube

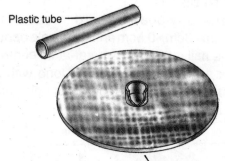

Plastic tube

Plastic disc with a hole

- Now, fill the balloon with air and put it on the tube. The air will pass through the fine bore of the disc and will provide an air cushion under it. Give a slight push to the disc with the tip of your finger. It will move. This model will work only on a smooth surface. ■■

24

12. Making a clinometer

A clinometer is an instrument used to measure the height of any building. It is an important instrument used by surveyors and civil engineers. It measures the angle of inclination with reference to a plumb bob. In this instrument a plumb bob or weight attached to a piece of string hangs from the sighting device. As the device is tilted upward to view the top of the building, the string moves across a protractor like face and indicates the angle of inclination. In this project you can make a simple clinometer and can measure the height of a building or of a tree.

You Require
- *One protractor*
- *A piece of thread*
- *One nail of about 1" length and a log table*

What To Do
- Take a protractor, a piece of thread and a nail. First make a tiny hole at the centre of the base line on the protractor. Pass the thread through the hole. Tie a knot in

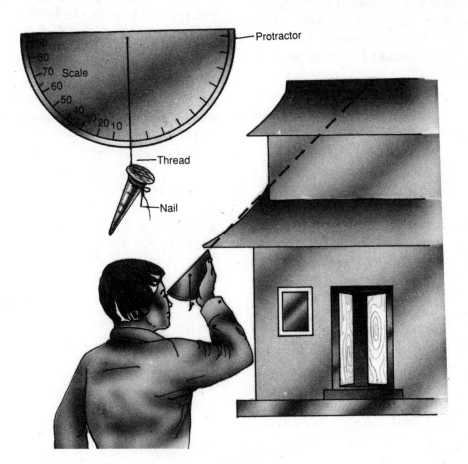

Measuring the height of a building

25

one end of the thread and tie the nail to the other end.

- When the protractor is held with its straight edge horizontal and upside, the nail should hang just below the curved edge at the mark of 90°, as shown in the figure.

The clinometer is now ready and can be used for finding the height of any building as follows.

Finding the height of a building

- Look along the top edge of the clinometer, and tilt it to point at the top of a building. Allow the thread to move freely around the scale, as shown in the figure.
- Press the thread against the scale, then read the angle through which it has moved. This will be equal to the angle which the building makes with the horizon.
- Now, measure the distance from the place where you stand to the base (bottom) of the building.

- The height of the building can now be calculated by multiplying this distance by the tangent of the angle you measured. Suppose h is the height of the building and d the distance between your place of measurement and the base of the building. And θ denotes the angle which you have measured. Then,

$$h = d \tan \theta$$

The value of tan θ can be seen from the trigonometric log tables.

Note: You can make another kind of simple clinometer out of a card board sheet cut to the shape of a right-angled triangle (45°, 45°, 90°). This will need a plumb live, and a measuring tape. By lining up the longest side of the triangular card with the top of a tree, or pole or building so that the plumb line hangs exactly along the short side, you can calculate the height of these objects. You will need to walk towards or away from the tree until this happens, Now measure the distance between yourself and the foot of the tree. Its height is that distance plus your own height.

■■

13. Making an automatic rain-gauge with time indicator

The rain-gauge is an instrument used to measure the rainfall. It is a very important instrument for weather forecasters. In this project, a rain-gauge for the measurement of rainfall, along with the time of raining, has been given.

You Require

- *One cylindrical plastic jar*
- *Plastic tubing*
- *A clock*
- *Graph paper*
- *One fountain pen*
- *One funnel*
- *One floating lever*

What To Do

- Take one cylindrical plastic jar and make two holes, one at the lower portion and other almost in the middle. Fit a plastic tube with a funnel in the lower hole and another plastic tube in the upper hole. This tube is provided to discharge the extra water in case of excessive rainfall.

- Put a floating indicator in the jar having a signature pen attached at the top. The pen should be in contact with the graph paper, as shown in the figure, to mark the amount of rainfall.

- The graph paper can be calibrated with a clock to ascertain the time.

■■

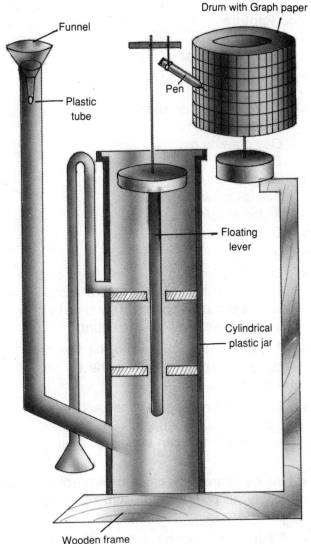

Funnel

Drum with Graph paper

Pen

Plastic tube

Floating lever

Cylindrical plastic jar

Wooden frame

14. Making an anemometer

An anemometer is an instrument used to measure the velocity of wind. The simplest anemometer has three or four cups attached to a vertical pipe. The wind catches the cups, spinning them around. The winds speed is measured by the number of times the cups go round in a certain period of time. Aeroplane pilots and sailors need to know the speed of the wind. A meteorologist makes use of anemometer for making weather forecasts.

You Require
- *One shoe box*
- *Four paper cups*
- *Stiff wire*
- *Tape*
- *Two thin strips of wood*
- *One ball-point pen*
- *One drawing pin*
- *One pin*
- *Thread*
- *A small roller*
- *A stop watch*
- *One small weight*

What To Do
- Take one foot length of stiff wire and bend one of its ends in the form of a loop. Fix this end at the bottom of the shoe box with tape, as shown in the figure.
- Put some stone pieces in the box to make it stable and place the lid on, making a hole for the wire.
- Take two thin pieces of wood and make a cross. Push four paper cups onto the ends of the wooden cross, facing in a circle.
- Fix the centre of the cross to the top of an empty ball-point pen tube with a drawing pin and tape.
- Push a pin through the tiny hole in the side of the tube and place it on the wire upright, so that it pivots easily on the pin.

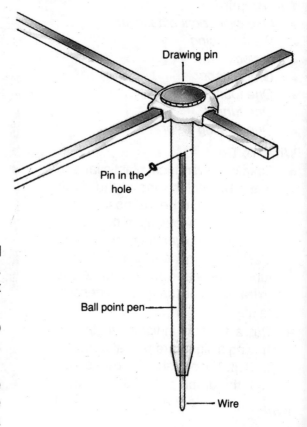

Drawing pin

Pin in the hole

Ball point pen

Wire

- Tie a cotton thread to the pin in the side of the tube, leading it sideways over a roller. Let it hang down at the side of the table as shown in the figure. Tie a small weight at the other end of the thread, just touching the ground.

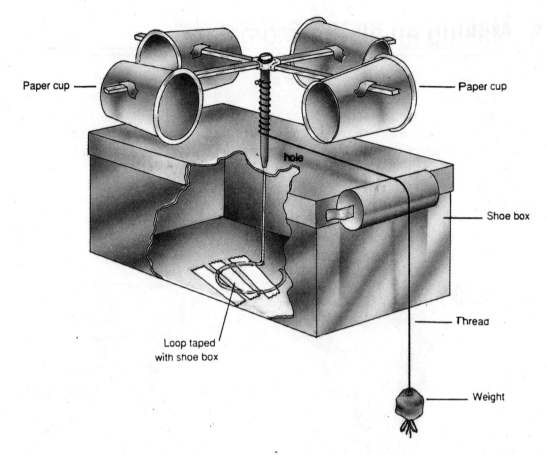

Paper cup

Paper cup

hole

Shoe box

Thread

Loop taped
with shoe box

Weight

- Now, put the whole arrangement in a windy place and let the cups go. They should swing round in the wind. Use a watch to note the time, how long the weight takes to rise from the ground to the edge of the table. Now, divide this length by the time taken, and you will get, wind velocity.

Note: You can make a portable anemometer as follows:

- Use a protractor and compass to draw a scale of a piece of thick card.
- Tape half a tennis ball to a strip of card and cut a window so you can read the scale as shown in the figure.
- Pin the strip to the card and mark the scale

Angle	80	60	40	20
km/hr	13	24	34	52

15. Making an air thermometer

Usually, thermometers make use of the expansion of a liquid (mercury or alcohol), to measure the temperature. But an air thermometer makes use of air expansion for temperature measurement. Due to the rise in temperature, the air expands, by which a liquid is pushed down in a tube. The tube which is calibrated with the markings of temperature scales, indicates the temperature at a particular moment.

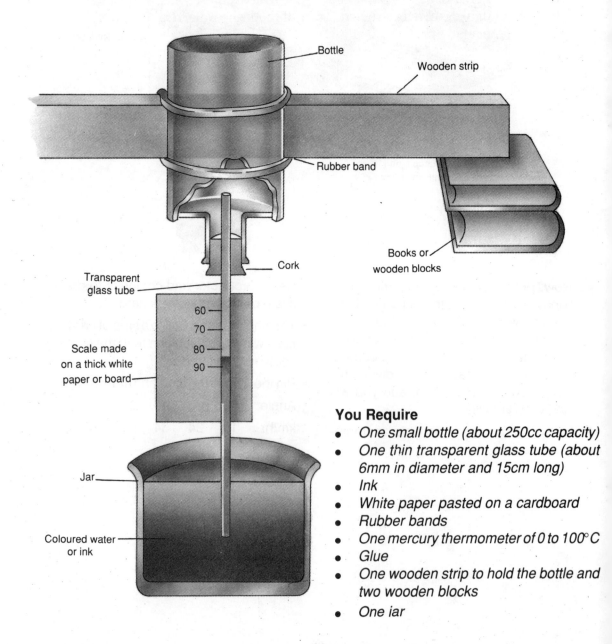

You Require

- *One small bottle (about 250cc capacity)*
- *One thin transparent glass tube (about 6mm in diameter and 15cm long)*
- *Ink*
- *White paper pasted on a cardboard*
- *Rubber bands*
- *One mercury thermometer of 0 to 100°C*
- *Glue*
- *One wooden strip to hold the bottle and two wooden blocks*
- *One iar*

What To Do

- Take a small bottle of about 250cc with a tight-fitting cork and a short thin glass tube.

- Make a hole through the cork with a nail and tightly fit the glass tube into the hole and push the cork into the neck of the bottle, as shown in the figure.

- Take a jar and pour some water into it and a little ink to make the water coloured. Place the jar on a table and support the bottle above it on wooden blocks and a strip of wood by rubber bands. Some part of the tube should dip into the coloured water of the jar, as shown in the figure.

- Glue the paper-pasted cardboard onto the tube.

- Warm the bottle with a hair drier to make the air expand. Some air will bubble out through the water. On cooling, the water from the jar will rise into the glass tube. Your air thermometer is ready.

- To calibrate this air thermometer, put the whole assembly in the sun along with the mercury thermometer. When the temperature is stable, mark a line at the liquid level on the paper-pasted cardboard and record the reading of temperature from mercury thermometer. Now, put the air thermometer in a cold room and mark another line on the paper-pasted cardboard. Divide the space between the two markings in equal parts and mark the scale. Now you can use this thermometer to measure the variations in temperature of the room during the day and night.

Note: Gas thermometers have the following advantages over the liquid thermometers:

(a) As the properties of a gas remain unchanged over a wide range of temperature, a gas thermometer is more suitable than a liquid thermometer for a wide range.

(b) As gases expand more than liquids, an air thermometer is more sensitive than liquid thermometers. ■■

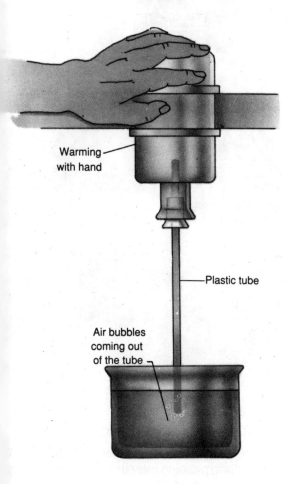

Warming with hand

Plastic tube

Air bubbles coming out of the tube

16. Making a wave machine

A wave machine is a simple machine used to study wave motion on the surface of water. By the rotation of a motor, waves are created on the surface of the water, and their motion is studied.

You Require

- *A small electric motor for 12 volt supply*
- *One battery of 9 volts*
- *A wooden board*
- *Screws or rubber bands*
- *An empty thread spool*
- *Small pieces of paper*
- *A short wooden rod with hook*
- *A string*
- *A cork*

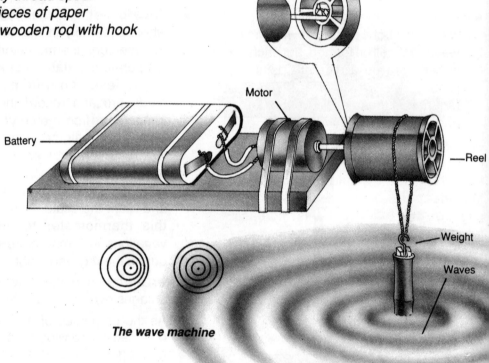

The wave machine

What To Do

- Take a small electric motor and connect it to a battery of 9 volt or even low.

- Fix the motor on the wooden platform with the help of the screws or rubber bands, as shown in the figure.

- Push the motor spindle into the hole of the empty thread spool. Tighten it with

the small pieces of paper.

- Hang the small wooden rod on a loop c string passing over the spool and throug the hook of the small wooden rod.

- Switch on the motor. The motor shoul run fairly slowly, making the spool wobbl and the rod bob up-and-down in wate

How to make waves and study them

- Take a bath tub or tray and fill it with water to half of its depth. Hold the wave machine over the tray so that the rod dips into the water. The up-and-down motion of the rod will produce circular waves on the surface of the water.

- By putting small pieces of paper on these waves, you can prove that the water does not move with the waves but only makes an up-and-down motion.

- Now, make the rod move through the water as it bobs up and down. Circles of waves will still be formed, but each circle is centered on a different point. As a result, the waves are compressed (squashed together) in front of the rod, and spread out behind it.

This wave pattern shows why the Doppler effect occurs. The waves pass any point in front of the rod at a greater rate, or frequency, than they pass a point behind the rod. In the case of sound waves, this increase in frequency corresponds to an increase in pitch. This is why a train whistle seems to have a higher pitch when the train is approaching.

A similar wave pattern also forms in front of fast-moving aircraft. At high speeds, the waves combine to form a shock wave. Try simulating this effect using a wave machine.

Note: The waves produced on the surface of water by a wave machine is probably the most familiar kind of waves, but there are many other kinds of waves all around us. For example sound is produced by waves. People's voices travel to our ears in waves. Sound waves are longitudinal waves. Radio and television programmes reach our homes in the form of electromagnetic waves. Light rays and X-rays travel in waves. These are electromagnetic waves.

Sound waves need material medium to travel while electromagnetic waves can travel even without any material medium. Waves of all kinds carry energy with them. All musical instruments produce waves of different frequencies.

17. Making a kaleidoscope

It is an optical instrument that forms beautiful patterns. It contains two mirrors which meet at an angle of 60°. Light from coloured objects placed near the mirrors bounces between them to form several reflections. The most fascinating aspect about the kaleidoscope is that you never get the same pattern twice. Designers of wall papers and fabrics make use of this instrument to get ideas for new patterns.

You Require

- *Two rectangular mirrors of the same size (8" long and 2" wide)*
- *One cardboard sheet of the same size as the mirrors*
- *Tape*
- *Tracing paper*
- *Transparent plastic paper*
- *Small pieces of bangles or coloured glasses.*
- *Glue*

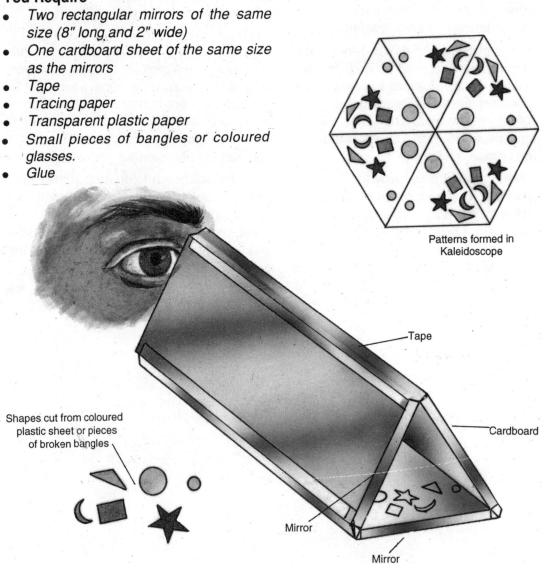

Patterns formed in
Kaleidoscope

Tape

Cardboard

Shapes cut from coloured
plastic sheet or pieces
of broken bangles

Mirror

Mirror

What To Do

- Take two rectangular strips of mirrors identical in size about 8 inches long and 2 inches wide. Take one cardboard piece of the same size as the mirror strips. Join the mirrors and the cardboard together to form a long triangular tube. The silvered sides of the mirrors should face inward. The angles between them will be 60°. This angle will produce six symmetrical patterns.

- Cut a piece of tracing paper or plastic paper to fit over one end of the triangular tube and glue or tape it in place.

- Take about 7-8 pieces of bangles of different colours. Put these coloured pieces in the tube. Then hold the tube upright so that the pieces rest on the tracing paper.

- Look down into the tube and see the pattern formed. The end of the triangular tube near the eye can be closed by a transparent plastic sheet or glass sheet.

The pattern will consist of 6 identical sections. Shake the kaleidoscope to change the patterns.

Alternate Method of Making a Kaleidoscope

- You can also make a kaleidoscope by using three plane mirrors of the same size and arranging them in a triangular pattern.

- Cover one end of these mirrors with tracing paper or ground glass plate and other with a card board. Make a hole in the card board to see through and drop pieces of coloured bangles or plastic inside. Point the end towards a light source. You will see beautiful patterns. When you turn the kaleidoscope, the coloured pieces change positions and new arrangements are formed.

- Do you know that kaleidoscope was invented in 1817 by Sir David Brewster, a Scottish physicist?

■■

18. Making a periscope

A periscope is an optical instrument with which a person can see around corners and other obstructions. It consists of a long tube with a reflecting mirror or prism at each end. The reflecting surfaces are parallel to one another. They are arranged at an angle of 45° inside the tube. Periscopes are important instruments used in submarines and tanks. They are also used by soldiers in the battlefield.

You Require

- *One long cardboard or plastic tube*
- *Two small mirrors*
- *One small hacksaw*
- *Glue*

What To Do

- Take a cardboard tube and two small mirrors. Use a small hacksaw to cut a section near each end of the tube. First make a horizontal cut to about 3/4 of the diameter of the tube. Now make two cuts at 45° to the horizontal cuts. Use a 45° set-square for guidance while making these cuts. The process is clear from the figure.

- Glue the mirrors in places where the sloping cuts have been made. The silvered surfaces of the mirror should face each other so that light can pass through the instrument as shown in the figure.

■■

Note: *Instead of using plane mirrors, you can use 90° prisms.*

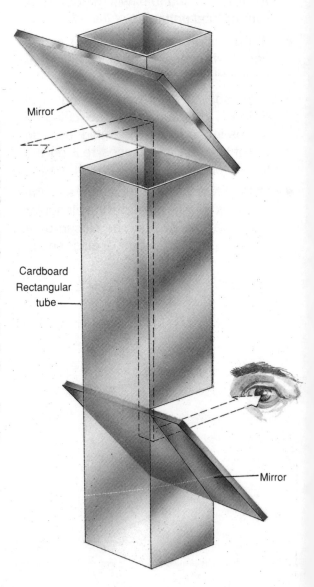

Mirror

Cardboard Rectangular tube

Mirror

19. Making a slide projector

A slide projector is an optical instrument used to see slides on a screen in an enlarged image. It makes use of a good quality magnifier and a light source. When light falls on the slide, its enlarged image is projected on a screen. This simple project will demonstrate the principle of projection.

You Require
- *One magnifying lens of about 50 cm focal length and 3" diameter*
- *Two cardboard boxes of the same size*
- *One piece of thick cardboard*
- *One milky bulb (about 60 watt)*
- *One ground-glass plate*
- *Glue*
- *Some slides*

What To Do
- Take one magnifying lens of about 3" diameter and 50 cm focal length.
- Take one cardboard box with a deep lid and of the same height as that of the focal length of the lens. The other dimensions of the box are not important.
- In the centre of the lid of the cardboard box, cut a hole to mount the lens. Then glue the lens in place.
- At the centre of the bottom of the box, cut a hole slightly larger than your slides.
- For the rear part of the projector, take the second box and make a hole in the bottom of this box of the same size as the hole in the front box. The two holes should line up with each other when the boxes are placed on their sides.

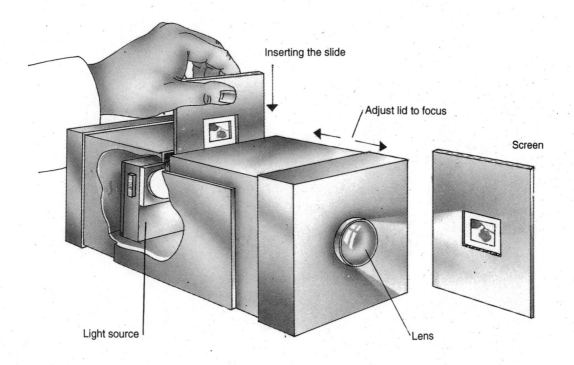

Inserting the slide

Adjust lid to focus

Screen

Light source

Lens

- Join the two boxes by gluing the pieces of cardboard to the sides. Leave a small gap between the boxes for the slide carrier.

- Make the slide holder from a piece of thick cardboard. This is used to position a slide between the holes in the boxes, as shown in the figure.

- Fit a milky bulb of about 60 watt in the rear box. Insert a slide upside down in the carrier and push this down into the projector. In a darkened room, project the slide on to a white paper or white wall by switching on the bulb. To get a sharp picture, adjust the focus by pulling out or pushing in the deep lid that holds the lens.

■■

Did you know ...

Slide shows existed even before the invention of photographic films. At that time, the projectors were called, "Magic Lanterns" because they appeared almost magical to the people who had never seen such things before. The projectors of that time consisted of a bright oil lamp inside a box, like a modern projection box but instead of film slides, hand painted glass plates were used as slides. Some slides were hinged to portray movement ; they would seem to show a dog wagging its fail, for example, or a ship sailing, as a handle on the slide was cranked.

20. Making an astronomical telescope

An astronomical telescope is an optical instrument used to see heavenly bodies like the moon and stars. It should never be used to see the sun because the powerful sunrays can cause severe eye damage. It makes use of two convex lenses — one small in diameter and the other of bigger diameter. The small lens is called the eyepiece and the bigger one is called objective lens. The separation between the two lenses should be equal to the sum of their focal lengths.

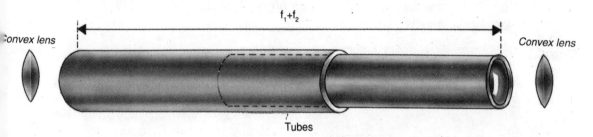

$f_1 + f_2$

Convex lens   Convex lens

Tubes

You Require

- *Two convex lenses — one with 2.5 cm aperture and 10 cm focal length and the other with 3.8 cm aperture and 50 cm focal length*
- *Two cardboard or rigid plastic tubes fitting into each other*
- *Glue*

What To Do

- Make the body of the telescope with two cardboard or rigid plastic tubes, one should slide smoothly into the other. The diameter of one tube should be slightly less than the diameter of the eyepiece and that of the other slightly less than the diameter of the objective lens. If the tubes do not fit tightly enough, wrap paper around the smaller diameter tube to make a better fit. The length of each tube should be about two-third of the distance required between the lenses to produce a sharp image.

- Take two convex lenses and glue them to the ends of the tubes. If the lenses are smaller than the ends of the tubes, mount the lenses in cardboard rings, as shown in the figure.

- Point the telescope at a distant object, and slide the tubes in and out until a sharp image is seen. The overall length in this situation will be equal to the sum of the focal lengths of the two lenses. You will see an inverted image of the object.

Note: To get an erect image of the object you have to use a right-angle prism between the two lenses.

Lens mount

21. Making a Galilean telescope

A Galilean telescope consists of a convex lens of larger aperture and larger focal length which acts as the objective lens, a concave lens of smaller aperture and smaller focal length which acts as the eyepiece. The two lenses are fitted at the ends of the two tubes which can slide into each other. The objective lens bends the light from the object into a clear, sharp image on the ocular lens. The eyepiece makes the image larger. The separation between the two lenses is adjusted to the difference of the focal lengths of the two lenses. It forms an erect image and is used to see terrestrial objects.

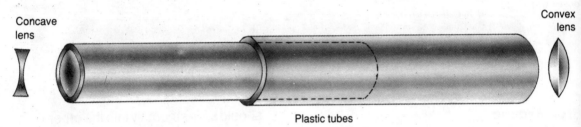

Concave lens

Convex lens

Plastic tubes

You Require

- *Two cardboard or plastic tubes fitting in each other*
- *One convex lens of 3.8 cm diameter and 50 cm focal length*
- *One concave lens of 2.5 cm diameter and 5 cm focal length*
- *Tape or glue*

What To Do

- Take one cardboard or plastic tube and fix the convex lens at its one end by tape or glue it.
- Similarly take another cardboard or plastic tube and fix the concave lens at its one end by tape or glue it.
- Insert the tube of smaller diameter into the tube of larger diameter and make them just able to slide by putting some paper lining.
- Now, point the objective lens of the telescope at a distant object. Focus by sliding the small tube in or out of the large one. You should be able to see a sharp, enlarged and upright erect image of the object.

■■

Warning

Never look directly at the sun through a telescope. Looking directly at the sun can cause eye damage.

22. Making a Newtonian reflecting telescope

A Newtonian reflecting telescope consists of a concave mirror. This mirror gathers light and focusses it on a flat mirror suspended in the centre of the telescope tube. The flat mirror reflects the light through a hole in the side of the telescope tube to the eyepiece lens and the object is clearly seen. You can make a simple newtonian reflector in this project.

You Require
- *One wooden or thick cardboard frame (as shown in the figure)*
- *One concave mirror (about 3" diameter and 50 cm focal length)*
- *Glue*
- *One plane mirror (about 1" × 1")*
- *One eyepiece*

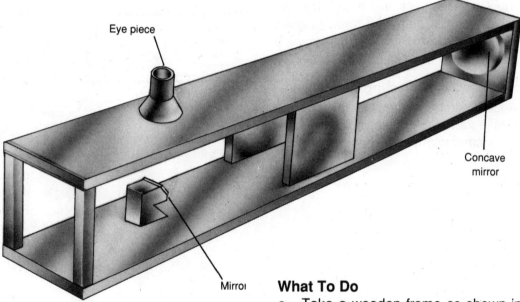

Eye piece

Concave mirror

Mirror

What To Do
- Take a wooden frame as shown in the figure (about 70 cm long, 4" wide and 4" deep). One of its ends should be open and the other end closed. At the top, there should be a hole to mount the eyepiece.
- Glue the concave mirror on the closed end of the wooden frame.
- At an angle of 45°, mount the flat mirror just below the hole meant for the eyepiece.

41

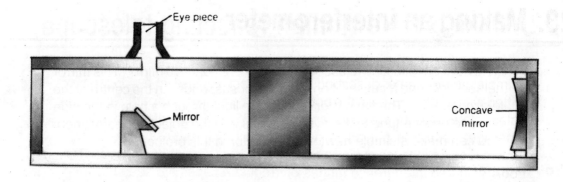

The separation between the concave mirror and flat mirror should be slightly less than the focal length of the concave mirror.

- Mount the eyepiece just above the frame as shown in the figure.

- View some object and adjust the 45° mirror slightly, back and forth, to get a clear view of the object. Your telescope is now ready to see distant objects.

Another kind of Reflecting Telescope

- Another kind of reflecting telescope is a Cassegranian telescope. In this telescope a concave mirror with a hole is used to collect light, Light reflected back by this mirror falls on a convex mirror which on further reflection falls on the eye piece. This produces an enlarged image of the object.

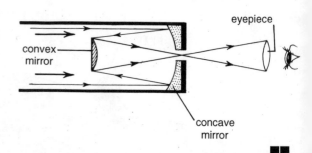

23. Making an interferometer

An interferometer is an instrument that produces interference of light. The simple interferometer described in this project, consists of a wooden box fitted with a light source at the bottom of the box in the rear part. A ground-glass plate is fitted in the box at an angle of 45° to the base of the box. When the light source is put on, it gets defused by the glass plate. If two microscopic slides are put in contact with each other in the front part of the box, coloured interference fringes of irregular shapes are seen between the slides. If the light source is a sodium lamp, the fringes will be yellow and dark in colour. Instead of two microscopic slides, if two optical flats are put in contact, straight fringes will be seen. This method is highly useful for examining the flatness of optical surfaces. Straight fringes indicate that the surfaces are flat and irregular fringes show that the surfaces are irregular.

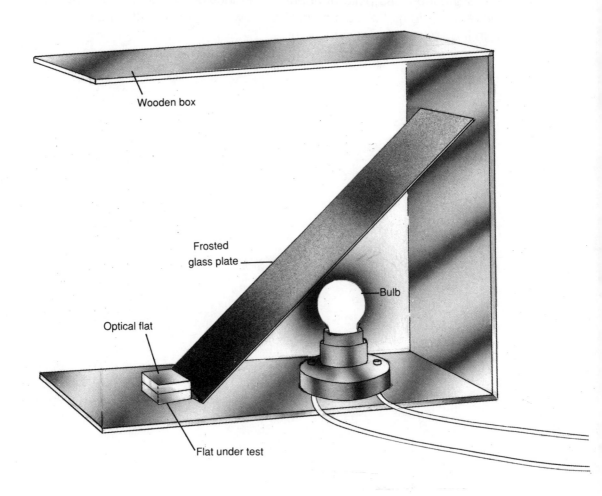

Wooden box

Frosted glass plate

Bulb

Optical flat

Flat under test

You Require

- *One wooden box (8" × 8" × 12")*
- *One milky bulb of 60 watt 220 volt*
- *A bulb holder*
- *One ground-glass plate*
- *Two microscopic slides*
- *One three-pin plug*
- *Plastic-coated wire*

What To Do

- Take one wooden box of about 8" × 8" at base and 12" in height. Its one side should be open. Fix a 60 watt electric bulb at the bottom of this box in the rear part, i.e. near the closed wall.

- Take a ground-glass plate of about 8" wide and 5.5" long. Fix it at an angle of 45° in the box, as shown in the figure.

- Now, put on the bulb and clean the two microscopic slides with a soft cloth. Put their surfaces in contact with each other in front of the inclined glass plate on the base of the box. Look into the slides carefully, you will see irregular coloured interference fringes.

This is your interferometer and you can use it to examine the flatness of any glass surface.

Note: This is a cheap interferometer and the cost involved is about 100 to 125 rupees. You can make a very good interferometer by using a sodium lamp of about 35 watts in place of the bulbs. In this case your fringes will be yellow and dark.

■■

24. Making a strain viewer

A simple strain viewer is a device by which the strains in glass and other transparent crystals are detected. It consists of a wooden box having an electric bulb fitted at the bottom of the box. Just above the bulb, a ground-glass plate is mounted to diffuse the light so that the illumination of the sample becomes uniform. Above the glass plate, a polariser is fitted in a wooden plate in the centre of the box. After leaving a space of about 4", another polariser (which acts as analyser) is mounted in the centre of the top. Two holes are made in the top and partition plates of the wooden box for mounting the polarisers.

To use the strain viewer, the polariser and analyser are adjusted in the cross position. In this position, no light is seen in the analyser. Now, the specimen, i.e. the glass plate is placed in between the polariser and analyser. The portions of strains in the glass plate appear as dark and bright patches due to the property of birefringence. These patches show the portions of strains in the glass plate. If a good glass plate without strain is put between the cross polariser and analyser, no such patch will be seen.

When the glass is made from the melt and allowed to cool a little quickly, the distances between the various molecules do not remain the same. As a result of this, the molecules are under stress, which give rise to strains. This defect can be removed by annealing the glass. Annealing is the process in which the solid is first heated to the softening point and then cooled down very slowly. Strain-free glass is used in the manufacture of optical components for precision optical instruments.

You Require

- *One wooden box with a partition in the middle*
- *One pair of sheet polarisers*
- *Ground-glass plate*
- *One bulb of 15 watt 220 volt with a holder*
- *Plastic-covered copper wire*
- *Three-pin plug*
- *One window glass piece*
- *One good quality glass plate*

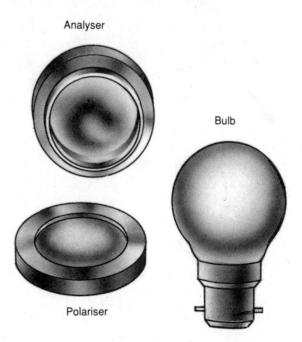

Analyser

Bulb

Polariser

Samples

What To Do

- Take a rectangular box of wood about 8"×8" base and 12" in height with a partition in the middle. Get two holes drilled in the centre of the top and partition plates equal in size to the diameter of the polariser and analyser.

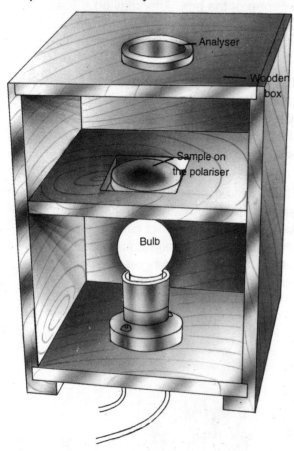

Stain viewer

- Fit the glass bulb at the bottom of the box in a bulb holder and take out its connections and connect a three-pin plug.

- Glue a ground-glass plate with the partition plate facing the electric bulb.

- Put the polariser in the partition plate hole and analyser in the hole at the top of the box. Cardboard disc with a hole can be glued under the analyser so that it does not fall.

- Connect the bulb to the AC mains and rotate the analyser to get the cross position of polariser and analyser. At the cross position, almost no light is seen coming out from the analyser.

- Now, put the sample on the polariser and view through the analyser to see strains in the glass plate. If coloured patches are seen, the glass plate has strains. Put another plate without strains—no such coloured patches will be seen, indicating the absence of strains.

Note: A Strain Viewer is a very useful optical instrument. It is used to examine the strains in glass by almost all optical industries. It is also used to examine strains in laser materials like ruby, neodymium YAG, Sapphire, fused silica, sodium chloride crystals etc.

■■

25. Making a polarimeter

A polarimeter is an optical instrument used to measure the specific rotation of an optically active solution, such as a sugar solution. It consists of a light source, a pair of polarisers and a tube for filling the sugar solution. A beam of light is allowed to fall on a polariser. At some distance, an analyser is mounted on a rotating device. The analyser and polariser are firstly set at cross position. When a tube filled with sugar solution is placed between the polariser and analyser, the direction of polarisation is rotated. When the analyser is rotated again, we get the position of extinction. The rotation of analyser is measured. This is the rotation which is caused by the sugar solution.

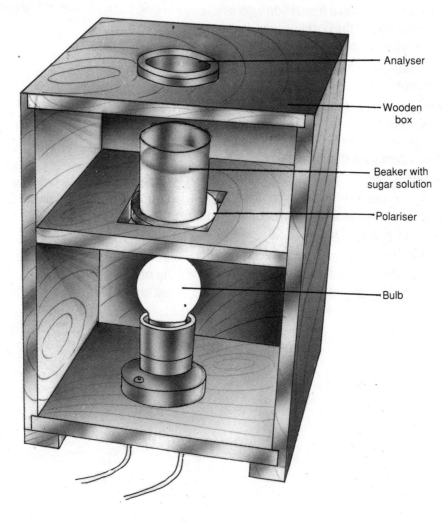

Analyser

Wooden box

Beaker with sugar solution

Polariser

Bulb

Polarimeter

You Require

- One pair of polarisers
- One wooden box having a partition
- One beaker
- One bulb
- One bulb holder
- Sugar
- Water connecting wire
- One circular scale

What To Do

- Get a wooden box made having a height of 12", and the base area of 8"x 6". Fit a bulb holder and bulb at the bottom of this box. It should have a partition with a hole in the centre of the partition wall. It should also have a hole at the top.

- Put one polariser in the hole of the partition wall.

- Put another polariser at the top of the box and fix the circular scale on the periphery of this polariser. Mask half of this polariser with a semicircular sheet of paper.

- Take a beaker and make the sugar solution in water. The solution should be quite concentrated.

- Rotate the analyser to get a cross position (when the light intensity is at minimum). Now put the beaker with sugar solution above the polariser at the partition wall. You will see the cross position is disturbed. Again, rotate the analyser to get the cross position. The angle with which you have rotated the analyser gives the rotation of polarisation by the sugar solution.

Note: From the angle of rotation of polarisation you can measure the specific rotation of the sugar solution. The specific rotation is defined as the rotation produced by a 10 cm long column of the liquid containing one gram of the active substance (sugar) in one cubic cm of the solution. Therefore

$$\alpha = \frac{100}{lc}$$

where α is specific rotation, θ is the angle of rotation, l is the length of the solution in centimeter through which plane polarised light passes and C is the concentration of the active substance in gm/cc in the solution.

You can also do the same experiment with camphor in alcohol because this solution is also optically active.

Polarimeter is a very important instrument. It is used in sugar mills to determine the quantity of sugar in a solution. In sugar mills this instrument is called saccharimeter.

■■

26. Making a direct vision spectroscope

A direct vision spectroscope is an optical instrument designed for the qualitative study of the spectra of some light sources in its line of sight. This is the most suitable instrument for educating students with various types of spectra produced by different light sources. It can also be used for the identification of various elements present in the light sources.

A direct vision spectroscope consists of five prisms cemented together. The two prisms in this combination are made of flint glass and three prisms are made of crown glass (as shown in the figure). When a collimated beam of sunlight or fluorescent tube light falls on these prisms through a slit, it splits into its constituent colours. These colours or spectra are seen through the eyepiece and focussed properly. We can see different emission lines contained in the light. We can also study Fraunhofer lines in the sun's spectrum by this instrument.

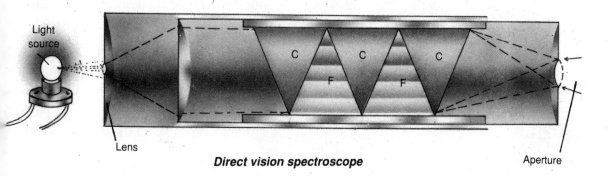

Direct vision spectroscope

You Require

- *Three 60° prisms made of crown glass*
- *Two 60° prisms made of flint glass*
- *One plano-convex lens*
- *One plastic tube about 6" long and of a diameter in which the prisms can be inserted in*
- *One slit made of aluminium foil*
- *One eyepiece*

What To Do

- Buy three crown glass prisms and two flint glass prism from the shop of a scientific dealer or optics dealer and get them cemented with each other such that flint glass prism lies in the centre.

Canada balsam is used for cementing the prisms.

- Fix the single slit and lens to the plastic tube with tape or adhesive.

- Now, insert the prism combination into the tube as shown in the figure and glue them at one edge.

- Put the eyepiece at the other end of the plastic tube and look through the whole assembly towards the fluorescent tube light. You will see many bright lines of different colours in the field of view. Now, adjust the eyepiece back and forth to get the well-defined and sharp spectral lines.

Your spectroscope is ready. You can study the spectrum of sunlight or bulb with it. In the sun's spectrum, you will see bright lines as well as some dark lines. The dark lines are Fraunhoffer lines.

Note: You can also make a direct vision spectroscope by using two crown glass and one flint glass prism.

Viewing the Spectra of Different Light Sources

With the help of this direct vision spectroscope you can study the spectra of different light sources.

- When you look at an incandescent light source (electric bulb) through this spectroscope, you will see a continuous strip of seven colours (VIBGYOR) without any line. This is known as continuous spectrum.

- When you see a sodium street light or a sodium lamp in your laboratory, you will observe two yellow lines in the spectrum. These are two emission lines of sodium.

- It you see the sun light through this spectroscope, you will see emission lines along with some dark absorption lines. These dark lines are called Fraunhofer lines.

- If you see the tube light through the direct vision spectroscope, you will observe several bright lines of different colours.

27. Producing electricity from potatoes

A primary voltaic cell makes use of a dilute solution of sulphuric acid in which one plate each of copper and zinc is dipped. The electricity in this cell is produced due to the migration of electrons from the zinc plate to copper plate in the dilute sulphuric acid solution. In fact, dilute sulphuric acid provides the path for the movement of electrons. In a similar way, potatoes act like the solution of dilute sulphuric acid between the copper and zinc strips which are used for producing electricity from potatoes.

You Require
- *8 medium-sized potatoes*
- *8 strips of copper 5 cm × 1cm*
- *8 strips of zinc 5 cm × 1cm*
- *Copper wire*
- *One 5 volt bulb with holder*

What To Do
- Take 8 medium-sized potatoes and insert a strip of zinc on one end of each potato. Similarly, insert one copper strip on another end of each potato, as shown in the figure.

- Arrange these potatoes on a table in two rows, each containing four potatoes. Connect the zinc strip of one potato to the copper strip of the other with the copper wire as shown in the figure.

Connect all zinc and copper strips with the help of the copper wire.

- Now, take one wire connected to the zinc strip and one wire from the copper strip and connect these two with the 4.5 volt bulb. You will see that the bulb immediately glows because the whole assembly acts as a battery. The bulb will glow only for a limited time.

Other suggested projects
- You can make similar batteries by using lemons instead of potatoes. The entire method will remain the same.

- You can make batteries by taking the juices of orange, lemon and other citrus fruits in beakers and dipping in them the zinc and copper strips.

■■

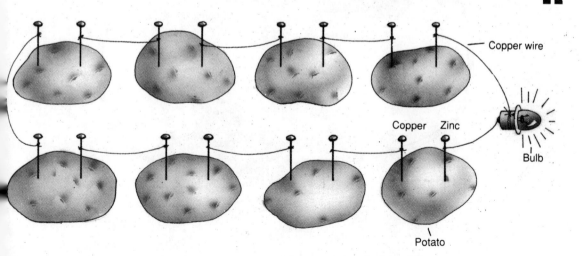

51

28. Making a dry cell

A dry cell is a portable battery used in torches, transistor radio sets, calculators, digital watches, etc. as a source of electricity. A conventional dry cell consists of a cylindrical vessel of zinc with a carbon rod in its centre. A paste of manganese dioxide and ammonium chloride is filled in it. When the zinc vessel and carbon rod are connected to a bulb by means of a wire, the bulb glows due to the flow of current. The voltage of such a cell is about 1.5 volt. You can make a model of such a dry cell as follows.

You Require

- *One zinc plate 4" × 2"*
- *One carbon plate 4" × 2"*
- *Manganese dioxide*
- *Starch powder*
- *Ammonium chloride*
- *Cotton wool*
- *Copper wire*
- *Two crocodile clips*
- *One 1.5 volt bulb and one bulb holder*

What To Do

- Take some water and mix starch in it Boil the mixture till the starch forms a thick paste. Mix sufficient quantity o manganese dioxide with the paste o starch, making a very thick paste c manganese dioxide.
- Take the zinc plate and uniformly spread the manganese dioxide paste over it.
- Take some cotton wool, spread it to the

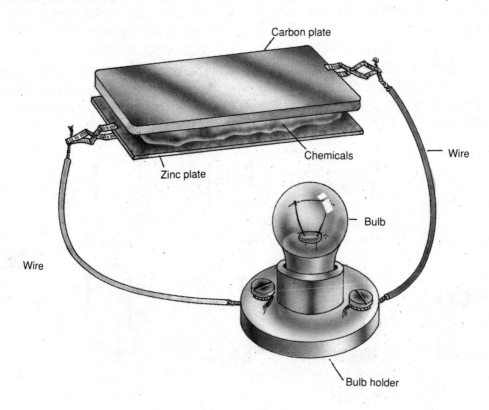

Carbon plate

Chemicals

Wire

Zinc plate

Bulb

Wire

Bulb holder

size of the zinc plate and soak it in the saturated solution of ammonium chloride. Put it over the layer of paste. Put another layer of manganese dioxide paste over it.

- Now, put a carbon plate on the above layer of manganese dioxide. Your dry cell is ready for use.

- Take two copper wires. Connect their two ends to two crocodile clips and other two ends to a bulb holder. Clip the zinc plate and carbon plate with the two crocodile clips separately. The electric current will flow through the bulb and the bulb will glow.

Note: There are three main types of dry cell batteries : Carbon-zinc, alkaline and mercury. The one described here is a carbon-zinc dry cell.

An alkaline dry cell battery is more powerful. It lasts five to eight times longer than a carbon-zinc battery. It has a carbon electrode and a zinc casing electrode. The electrolyte is a strong alkali solution, potassium hydroxide. Alkaline dry cells are used mainly for portable radios.

In a mercury dry cell, the voltage remains constant to the end of the battery's life. These are used in toy cars, electric razors, deaf aids etc. Electric toothbrushes often use this type of cells. In this cell, a mercuric oxide electrode is used. The other electrode is the zinc casing. The electrolyte is potassium hydroxide.

Nickel cadmium batteries though wet cell batteries can be sealed air tight and are ideal for use in portable tools and equipments.

■■

29. Converting solar energy into electrical

The solar energy is converted into electrical energy with the help of a solar cell. A solar cell is a silicon wafer that converts light energy into electrical energy by means of the photoelectric effect. The amount of electrical energy produced by a solar cell can be greatly increased by using a concave reflector to focus the sunrays.

You Require

- *One solar cell (it can be made by removing the metal cap of a power transistor)*
- *One milliammeter.*
- *One cardboard box*
- *2 pieces of insulated wire*
- *One concave mirror*

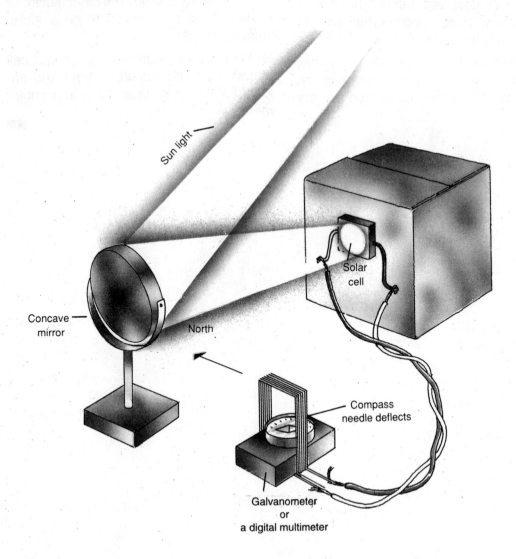

Sun light

Concave mirror

North

Solar cell

Compass needle deflects

Galvanometer
or
a digital multimeter

What To Do

- Purchase a power transistor and remove its metal cap to make a silicon solar cell. Take one milliammeter (0 to 10 milliampere range) from an electronic dealer.

- Glue the back of the solar cell onto a cardboard box as a support.

- Connect two pieces of insulated wire of the cell to the ends of the milliammeter.

- Shield the cell from light. Your ammeter will show zero deflection.

- Now allow the sunlight to fall on the cell. The pointer of the ammeter will show a small deflection.

- Now, use a concave mirror to reflect the sunrays onto the solar cell. The mirror focusses the sunlight onto the cell. In this case, the ammeter will show greater and greater deflection as the size of the patch of light on solar cell is reduced by moving the mirror back and forth. It indicates that the photoelectric current increases as the amount of light falling on the solar cell increases.

Note: It is a small project demonstrating the conversion of solar energy into electrical energy. Now-a-days solar cells, each capable of producing about one half of a volt of electricity, are available. To produce useful quantities of electric current, it is necessary to join great numbers of cells. The cells are joined together in big panels and are placed in sun light. The electricity produced is used to charge electric storage batteries.

Solar cells are being used to provide booster current for telephone cables. Solar cell panels are being used in space crafts, space laboratories and satellites to provide power for their scientific equipments and transmitters. Now-a-days solar cell operated calculators and radios are also available.

■■

30. Making traffic lights

Traffic lights are used in big cities at major road crossings to make the vehicular movement smooth and avoid accidents. The red light is a signal of 'stop', green light is a signal of 'go' and orange light is a signal of 'ready'. These lights are operated automatically. You can make a model of the traffic lights.

You Require

- *Two wooden strips — one 12" × 4" × 1" and the other 8" × 4" × 1"*
- *Three bulbs — each 3 volt (one red, one green, one orange)*
- *Three bulb holders*
- *Enamelled copper wire*
- *Two dry cells each 1.5 volt*
- *4 drawing pins*
- *One U-clip*
- *Two wood screws*

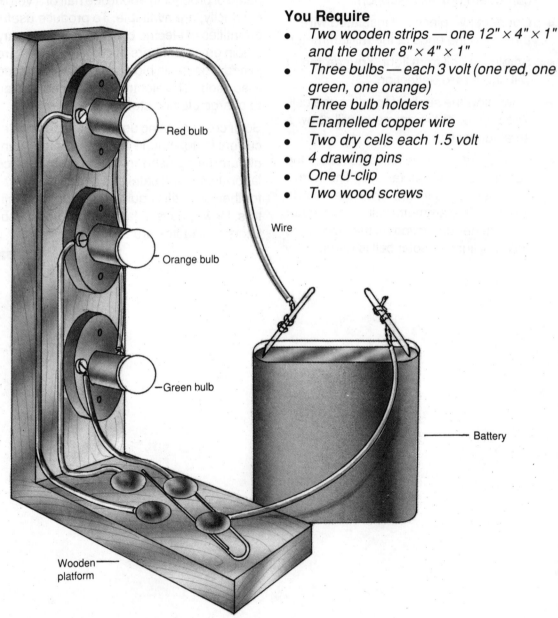

Red bulb

Orange bulb

Wire

Green bulb

Battery

Wooden platform

What To Do

- Take two wooden strips—one 12" × 4" × 1" and the other 8" × 4" × 1". Put them at right angles and fix them with the help of wood screws as shown in the figure.

- Put three bulb holders in a line and wire them in parallel. Insert the bulbs in the holders: top — red, middle — orange and bottom — green.

- Fix the three wires from three bulbs at the base with drawing pins and put a fourth pin by pressing a U-clip under it. One end of it should be straightened so that it can be pressed against either of the three pins. This will act as a switch.

- Connect the two dry cells in series, i.e. the positive terminal to negative of the other, to each bulb.

- Connect the common wire from the bulbs to the positive terminal of the dry cell combination. Connect one wire from the negative terminal of the dry cell combination to the U-clip. Your model is ready.

- Now, when you press the straightened edge of the clip against any of the three pins, one of the three bulbs will glow.

Note: You can make a simpler model of traffic lights by using only two bulbs—red and green instead of three, on the same lines as given above.

Did you know ....

Many countries have developed robots for controlling traffic. Such robots are being used in Europe and USA. Silent Sam is one such robot that can control traffic like a traffic police man.

31. Making an electromagnet

An electromagnet consists of an iron core over which enamelled copper wire is wound around. When electric current is passed through the coil, the iron core becomes a magnet. This is known as an electromagnet. The magnetic strength depends upon the number of turns of the coil and the current flowing through them. As the number of turns is increased, the strength of the magnetic field also increases. Similarly, the strength of magnetic field also increases with the increase of electric current. You can measure the strength of magnetic field in this project by changing the number of turns but keeping the voltage constant. The strength of magnetic field can be measured by using a spring balance as follows.

You Require
- *One iron rod or a bolt 4" long and half inch in diameter*
- *Two dry cells connected in series to give a voltage of 3 volt*
- *Enamelled copper wire of 40 gauge*
- *One stand*
- *One small spring balance or a compass*
- *Tape*

What To Do
- Take the iron rod or a bolt and make four windings of copper wire over it with 150, 250, 350 and 450 turns. Each coil should

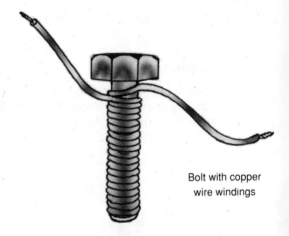

Bolt with copper wire windings

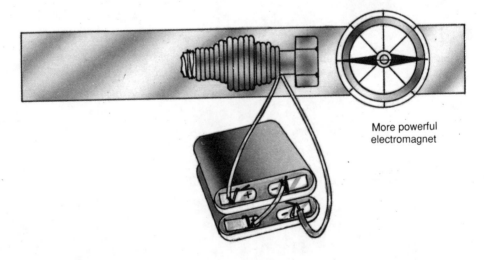

More powerful electromagnet

have free endings. Each layer of turns can be wound over each other with a paper wrapping in between.

- Hang the iron rod on a stand as shown in the figure. Connect the terminals of the coil of 150 turns to the two dry cells. The iron core will become a magnet.

- Bring a spring balance near to the lower end of the iron rod. It will be attracted and will stick with the lower end of the rod or put a compass on a ruler. It will be deflected by the electromagnet.

- Repeat the same process with the windings of 250, 350 and 450 turns. You will see that each time the weight required for detaching the balance becomes more and more. If you are doing the experiment with compass, deflection will be more and more each time. This shows that the magnetic strength increases as the number of turns increases.

- Pull the balance gently till it gets detached from the rod. Note the reading of the balance.

- Now you can study the effect of electric current on the magnetic field by using only one dry cell. Connect the terminals of the coils of 150, 250, 350 and 450 turns, one by one to only one dry cell. You will see that the deflection of magnetic needle with one cell each time is less than the deflection obtained with two cells. This shows that the magnetic strength has decreased with the decrease of current flowing in the coils.

Note: Electromagnets are used in loud-speakers, electric bells, magnetic cranes, electric motors and generators. A large electromagnet can be used to remove steel splinters from the workman's eye in a workshop.

■■

32. Making a magnetic crane

A magnetic crane works on the principle of electromagnetism. Giant cranes are used to move heavy iron and steel scrap from one place to another. This project describes how to make an advance model of a magnetic crane.

You Require
- *Enamelled copper wire with bare ends*
- *Two pieces of thick cardboard 4 cm × 30 cm*
- *Scissors*
- *One round pencil*
- *Tape*
- *Two paper clips (U-pins)*
- *String*
- *One iron bolt about 1" long*
- *One cardboard box open at one end*
- *Two paper fasteners*
- *One cotton reel*
- *One 4.5 volt battery*

What To Do
- Wind about 50 turns of enamelled copper wire on the bolt. Leave two long pieces of wire free at each end (Fig. 1).
- Now, make small holes near each end of both the cardboard pieces and fix them inside the box with the help of paper fasteners as shown in Fig. 2.
- Fit the cotton reel between the cardboard pieces. To do this, straighten out a paper clip and push it through the holes in the cardboard and through the centre of the cotton reel. Bend down the paper clip ends to secure the cotton reel.
- Make a handle by attaching a bent paper clip to the end of the pencil with tape, as shown in Fig. 4.
- Make two holes in the cardboard box and then push the pencil through these two holes, as shown in Fig. 5.
- Tie one end of the string around the electromagnet. Lay the string over the cotton reel and tie the free end to the pencil. By turning the pencil, the electromagnet should move up and down Fig. 6.

Fig. 1 Electromagnet

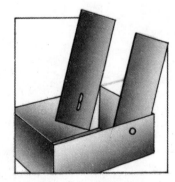

Fig. 2 Cardboard box

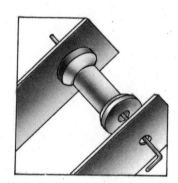

Fig. 3 Cotton reel

Fig. 4 Handle

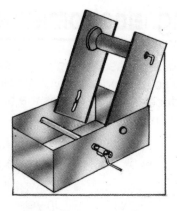

Fig. 5 Fitting the handle

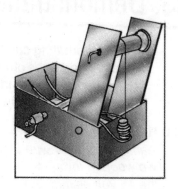

Fig. 6 Mounting electromagnet

● Now, connect the free ends of the electromagnet through a switch to the battery. Your crane is ready for use.

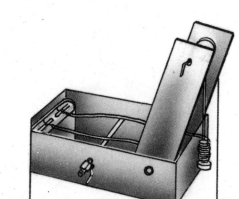

Fig. 7 Box on a table

● To use the crane, put some very small pieces of iron below the electromagnet. As soon as you switch it on, the bolt will become magnetic and iron pieces will cling with it. You turn the pencil, the pieces will be lifted up. When you wish to drop these iron pieces, switch off the electro-magnet, and the pieces will drop down.

■■

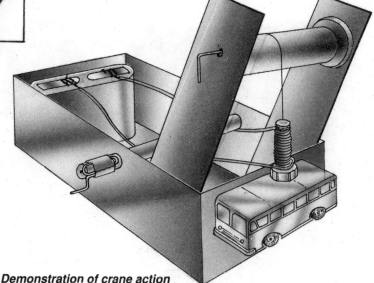

Demonstration of crane action

33. Demonstrating electromagnetic induction

When alternating current is passed through one coil, the induced voltage is developed in the other coil placed near the first coil. If a bulb is connected to the other coil, the induced voltage makes the bulb glow. This effect is known as mutual induction.

You Require

- *Enamelled copper wire 30 gauge*
- *Two wooden formers*
- *One 16 volt electric bulb*
- *One bulb holder*

What To Do

- Take one wooden former and wind a coil of enamelled copper wire of 30 gauge of about 1200 turns.

- Take another former and wind a coil of about 50 turns. Connect a bulb of 16 volt with the two free terminals of this coil through a bulb holder.

- Put the two coils near each other (about 2 cm apart), as shown in the figure.

- Now, put the two free ends of the first coil in your AC mains supply. Your bulb will glow. Bring the second coil with the bulb a little nearer to the first coil. It will glow more intensely. The bulb is lighted due to mutual induction.

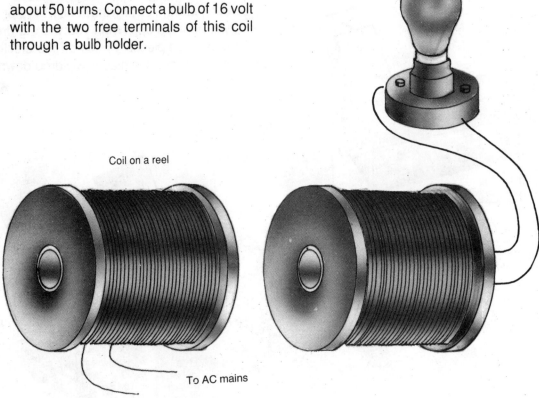

Bulb

Coil on a reel

To AC mains

34. Demonstrating spectacular levitation

The spectacular levitation can be demonstrated by allowing A.C. current to flow through a coil wound around a rod of iron and putting a ring over the rod. As soon as A.C. current flows through the coil, the expanding magnetic field induces an oppositely directed current in the ring. The field set up by the metal ring opposes the field of the coil by which the ring is thrown upwards in the air demonstrating the phenomenon of magnetic levitation.

You Require

- One rod of iron about 12 cm long and 1 cm diameter
- One metal ring having a hole of about 1.5 cm dia.
- Two circular cardboard pieces
- Enamelled copper wire about 26 gauge

What To Do

- Put the two cardboard pieces on the iron rod separated about 8 cm from each other.
- Make a coil of wire between these two pieces at least with 400 number of turns.
- Now slip the ring on to the rod as shown in the figure.
- Connect the two terminals of the coil to A.C. mains. The ring will be thrown upwards into the air, demonstrating the phenomenon of magnetic levitation.

Note: Magnetic levitation is being used to lift and propel the trains.

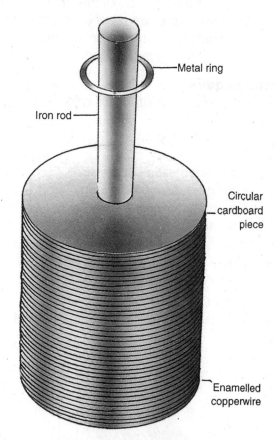

Metal ring

Iron rod

Circular cardboard piece

Enamelled copperwire

35. Making an electric motor

An electric motor works on the principle of electromagnetic induction. An electric current flows into the motor through a wire. The wire inside an electric motor is wound into a coil that can rotate. This coil of wire is placed between two poles of a fixed magnet.

When an electric current flows through the wire, it creates a magnetic field around the coil, with a north pole at one side and a south pole on the other. If the south pole of the coil is near the south pole of the fixed magnet, the two like poles repel each other and the coil turns round. The coil rotates halfway round the circle. For a complete turning, the direction of the current in the coil is changed.

You Require
- *One wooden platform 15 cm × 10 cm × 1 cm*
- *One 5 cm long cork*
- *Two U-clips*
- *Tape enamelled copper wire*
- *Four nails about 5 cm long*
- *One hammer two drawing pins*
- *One steel knitting needle about 10 cm long*
- *One 4.5 volt battery*
- *Two long bar magnets*

What To Do
- Take a cork and push a knitting needle through the centre of it lengthways. Stick two pins into one end of the cork. They should be at the same distance from the needle. About one centimeter of each pin should be protruding out of the cork.

- Wind the coil of the enamelled copper wire about 50 turns lengthwise on the cork. Use the sticky tape to hold the coil in place. Scrap the ends of the copper wire.

- Wrap the bare wires around one of the pins and the other end to the other pin.

- Take the wooden platform and hammer

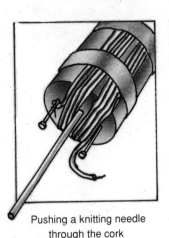

Pushing a knitting needle through the cork

Putting cross nails under the needle

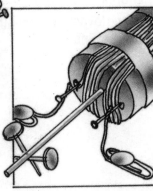

Making connections of the wire

two nails into each of its ends in such a way that they make a cross (X). Place the needle of the cork to sit on the cross. See that the bottom of the cork does not touch the wood base.

- Pull out one end of the paper clip and straighten it. Fasten the paper clip to the board with a drawing pin. When the cork is rotated, the pins on the cork should touch the straight end of the paper clip. Do the same thing with the second paper clip, placing it on the other end of the needle.

- Now, connect the battery to the paper clips by means of two small wires. Place one magnet at the side of the cork with the north pole closer to the cork. Place the second magnet on the other side of the cork with the south pole of the magnet closer to the cork.

- Rotate the cork slightly and your motor will start.

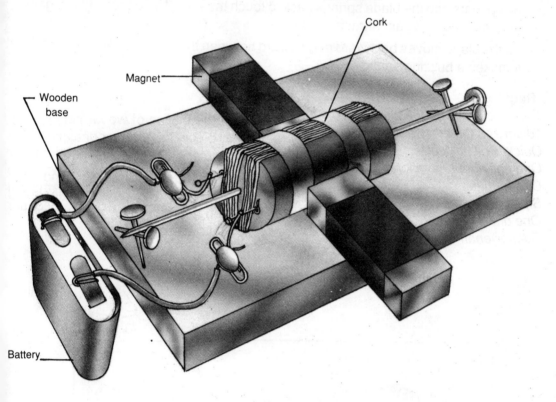

Note: The motor given in this project is not self starting and to start it the cork must be rotated.

36. Making an electric buzzer

The magnetic effect of electric currents makes an electric buzzer produce sound. They are used in doorbells. You can make a simple buzzer as described below.

When the bare wire from the electromagnet touches the blade, an electric current flows through both the wire and the electromagnet. The electric current creates a magnetic field around the electromagnet and so the steel blade is attracted towards the electromagnet.

When the blade moves towards the electromagnet and away from the bare wire, the electric current stops flowing. The magnetic field around the electromagnet disappears and the blade springs back to touch the wire. The current flows again and the magnet again attracts the blade.

As the blade moves backward and forward between the wire and electromagnet, it makes a buzzing sound.

You Require
- *Two pieces of wood—one about 6 cm × 10 cm and the other about 2 cm × 4 cm*
- *Quickfix or Fevicol*
- *One nail about 4 cm long*
- *One hammer*
- *Small piece of cloth*
- *One drawing pin*
- *Copper-enamelled wire with bare ends*

What To Do
- Take the pieces of wood and glue the small piece upright near one end of the larger piece as shown in figure.
- Hammer the nail into the other end of the wooden base. The top of the nail should be about 3 mm below the top of the upright piece of the wood.

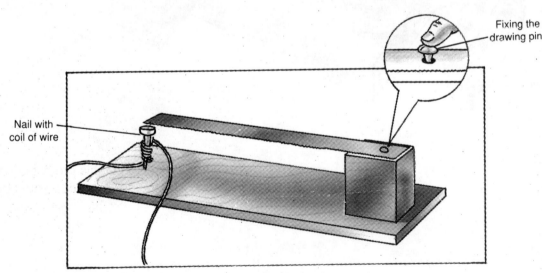

Fixing the drawing pin

Nail with coil of wire

Fig. 1

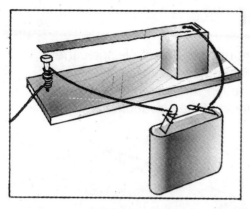

Connection plan

- Wind the copper wire around the nail (about 150 turns) to make an electromagnet. Leave about 30 cm of wire free at each end.

- Wrap the cloth around the saw blade and snap the blade into half. Handle the blade carefully to avoid cuts on your hands. If there is any paint on the blade, scrape it off with a piece of sandpaper.

- Now, push the drawing pin through the hole in the blade and fix it on the small wooden base, as shown in the figure. The other end of the blade should be about 3 mm above the nail.

- Connect one of the wires from the electromagnet to one terminal of the battery. Connect another terminal of the battery to the drawing pin through a small wire.

- Now, hold the wire from the electromagnet and gently touch the top of the blade with it. The blade will make a buzzing sound.

■ ■

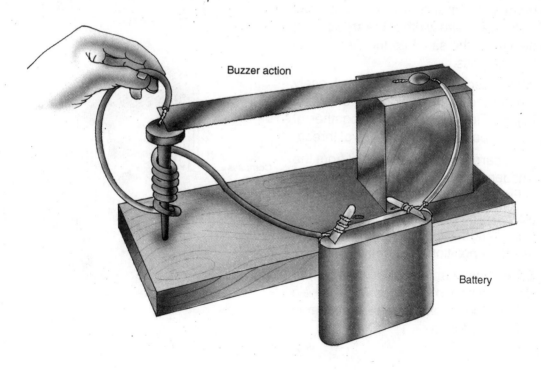

Buzzer action

Battery

37. Making a galvanometer

A galvanometer is an electrical instrument used to detect electric current. This instrument works on the principle of electromagnetism to detect electric currents. Even a weak current flowing through the instruments will show appreciable deflection in the magnetic needle of the galvanometer.

You Require

- *A pocket magnetic compass*
- *18 gauge enamelled copper wire about 10 m length*
- *A small block of wood*
- *Thread, sandpaper*
- *9 volt battery*
- *A small bulb and a bulb holder*

What To Do

- Saw a piece of a strip of wood about 4 × 2 inches to make the base of the instrument. The width of the strip should be almost the same as the diameter of the compass.

- Wind a rectangular coil of wire loosely around the block of wood as shown in Fig. A. Tie these windings together at various points with the pieces of thread.

- Now, carefully remove the coil from the wooden block and reposition it as shown in Fig. B. Glue it with the block.

- Now, fix the compass to the block with Quickfix so that the north pole of magnetic needle is positioned as shown in Fig. B.

- Scrape the enamel insulation of copper wire from its two free ends with the help of the sandpaper. The galvanometer is now ready for use.

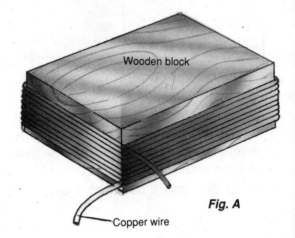

Wooden block

Copper wire

Fig. A

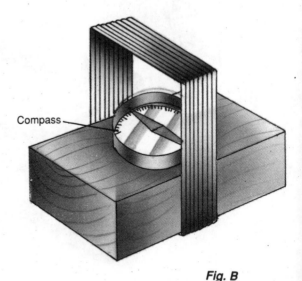

Compass

Fig. B

38. Detecting current using the galvahometer

- To detect the current with this galvanometer, you have to turn the instrument in such a way that magnetic needle directly lies under the coil, as shown in Fig. C.

- Connect the coil to the battery through a bulb, as shown in Fig. D. The coil will now act as an electromagnet and make the magnetic needle deflect to one side. On reversing the battery connections, the needle will deflect to the other side, as shown in Fig. E. The bulb is used as a resistance in the circuit to reduce the current flow and thus enable the battery to last longer.

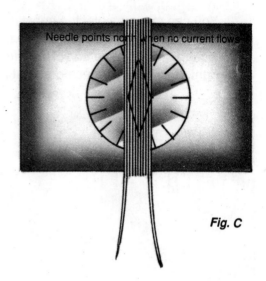

Needle points north when no current flows

Fig. C

■■

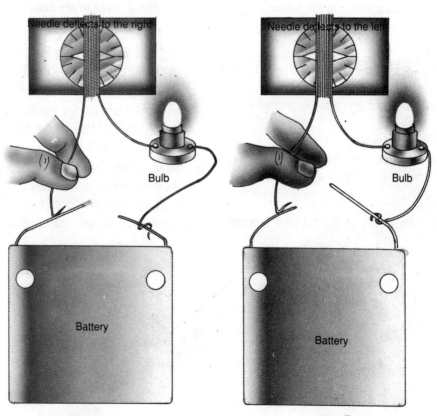

Needle deflects to the right

Bulb

Battery

Fig. D

Needle deflects to the left

Bulb

Battery

Fig. E

39. Making an electric board of birds' questions

You can make an electric question board to tell the name of a bird in one of the pictures. You have to touch one end of a wire to the screw tip under a bird picture and the other end of the wire to the screw tip under the name you think right. If you have selected a right name, the lamp will light.

Fig. 1 Drilling holes in the board

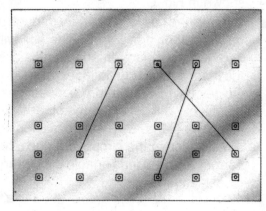

Fig. 2 Back of the question board

You Require

- *One thin wooden board (12" × 8"), 24 screws and bolt (1" size)*
- *12 pictures of different birds along with their names*
- *Plastic-coated wire*
- *Two dry cells connected in series*
- *One 3 volt bulb fitted in a bulb holder*

What To Do

- Take a thin wooden board of the size 12"×8". Mark places for the bird pictures and for the name cards. Glue them at the places as shown in Fig. 4

- Make a hole under each card and each bird. Screw a nut on each bolt as shown in Fig. 1 and 2.

- Now, at the back of the board, fasten a wire from the bolt under a picture to the bolt under the right name. Do this for each picture and card. (Fig. 2 & 3).

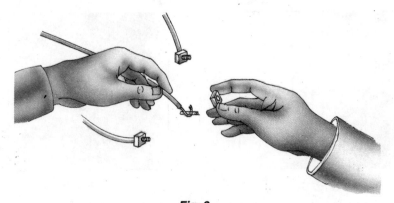

Fig. 3

- Now, connect a bulb through two dry cells as shown in the figure. Scrape the ends of the two wires. (Fig. 4.)

- Now, touch one end of the wire to the bolt under the bird and the other end to the bolt under the name card which you have thought as the correct name. If your answer is right, the bulb will glow.

In fact, the wire on the back of the board has completed the circuit of the bulb, that is why it glows.

Note: You can make such electric question boards by using pictures of flowers and their names, pictures of animals and their names, pictures of famous buildings and their names, etc.

■ ■

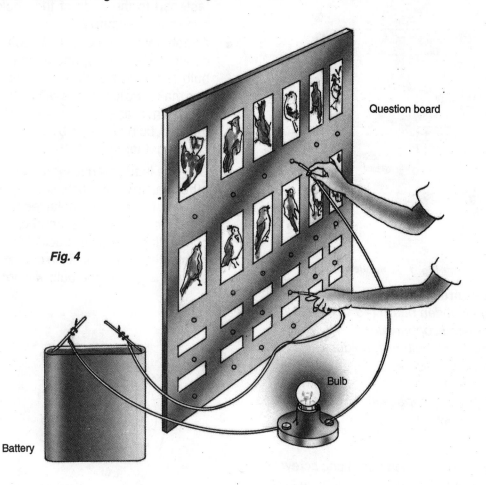

Fig. 4

Question board

Bulb

Battery

71

40. Making an electric quiz board

An electric quiz board consists of ten questions and their answers written on 20 separate strips of paper. They are randomly pasted on a wooden board. The backs of these questions and answers are connected with different copper wires. When the contact of a question is made with the right answer with the help of two wires, a bulb glows, showing that this is the right answer.

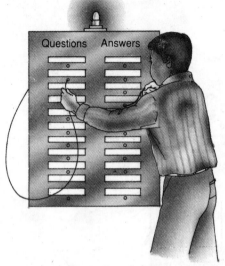

Fig. 1 Front view

You Require

- *One wooden board (45 cm × 15 cm)*
- *Insulated copper wire*
- *One 3 volt bulb with holder*
- *Two dry cells*
- *20 cup hooks*
- *20 strips of paper with 10 questions and their answers*

What To Do

- Take one wooden board and screw two lines of metal hooks through the board.

- Paste down your questions against one line and the answers, jumbled up, against the second line. (Fig. 1)

- Connect each question to the correct answer by a wire at the back of the board fastened to the ends of the hooks, as shown in the figure 2.

- Attach two dry cells (1.5 volt each connected in series) and a small 3 volt bulb to the board. Wire them together and make two long leads with metal clips on the ends, as shown in the figure. Test that the bulb will light by touching the clips together. (Fig. 2)

- Clip one lead to a question and clip the other lead to what you think is the correct answer. If the answer selected by you is right, the bulb will light up because the connecting wire on the back of the board completes the circuit. If the choice of the answer is wrong, the bulb will not light. ■■

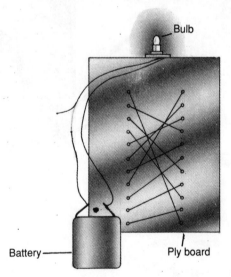

Fig. 2 Back of the question board

41. Making a railway signal

Railway signals, just like traffic lights on roads, are used to give indication to make trains stop or move. When a signal is down, it indicates a clear path to the train. You can make a working model of a railway signal in the following way.

You Require

- *Two pieces of wood—one (10 cm×2 cm) and the other (5 cm × 2 cm)*
- *Fevicol*
- *One small piece of thin cardboard*
- *Two pins*
- *Cotton thread*
- *A sewing needle*
- *Enamelled copper wire*
- *Modelling clay (plasticin) or Quickfix*
- *One switch*
- *One iron nail*
- *One piece of drinking straw (about 3 cm long)*
- *One battery of 4.5 volt*

What To Do

- Fix the long piece of wood on the short piece in an upright position with Fevicol or wood glue.
- Cut the signal arm from the cardboard about 5 mm wide and 5 cm long. Colour it with water colours.
- Use a pin to fix the signal arm to the upright piece of wood near the top. Place the second pin just beneath the signal arm so that the signal rests on the top of it in a horizontal position. (Fig. 1)
- Use a needle to make a hole near the bottom of one end of the signal arm. Tie a piece of cotton thread through the hole. Cut the thread so that it reaches till just above the wooden platform. Tie the other end of the thread round the nail and leave it hanging. (Fig. 2 & 3)
- Take the drinking straw and wind about 500 turns of copper wire over it. Leave about 25 cm of wire free at each end. (Fig. 4)
- Attach this straw solenoid to the wood base with Quickfix as shown in the figure 5.

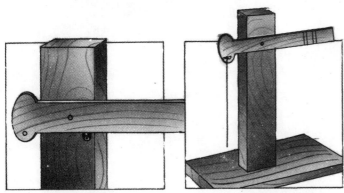

Fig. 1 Signal with wooden board

Fig. 2 Cotton thread with signal

Fig. 3 Knot in the thread

Fig. 4 Coil of wire

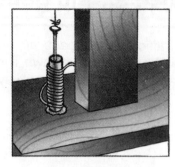

**Fig. 5 Fixing the coil
with wooden base**

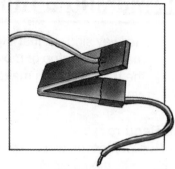

Fig. 6 Switch

- Take a switch and connect the two free ends of copper to it through a battery, as shown in figure 6.

- When you turn the switch on, an electric current flows through the circuit. The current creates a magnetic field that pulls the nail into the centre of the coil, which in turn makes the signal arm go up. When the switch is put off, the nail comes out of the coil and the signal arm falls down.

■ ■

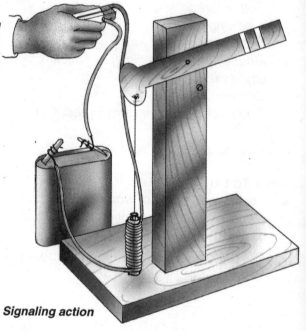

Signaling action

74

42. Making a thermoelectric generator

When two dissimilar metals are joined and one junction is heated while the other is kept at low temperature, an electric current flows through the wires. The voltage so developed is called thermo e.m.f. The direction or flow of current depends upon the combination of metals used. For example, in a combination of copper and iron, the direction of the current flow is from copper to iron at the hot junction. Let us make a thermoelectric generator and demonstrate the production of electricity.

You Require
- One milliammeter
- Copper wire
- Iron wire
- Beaker
- Ice
- Bunsen burner or candle

What To Do
- Take one copper wire about 60-70 cm long and one iron wire of the same length. Join the two ends of the copper wire with the two ends of iron wire.

- Cut the copper wire at the middle and join its two ends with a milliammeter, as shown in the figure.

- Now, heat one junction of copper and iron wires with a Bunsen burner or candle and put the other junction in the beaker. Put some ice pieces in the beaker.

- After the junction becomes hot, you will observe that some current flows in the milliammeter. This is the thermoelectric current.

■■

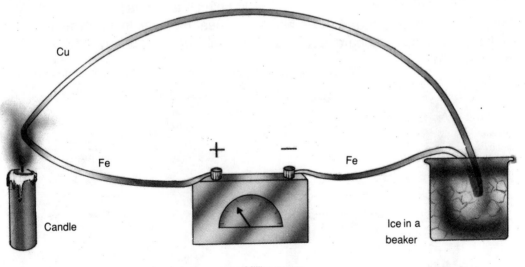

Cu

Fe   +   −   Fe

Candle

Milliammeter

Ice in a beaker

43. Making a chimney for controlling smoke pollution

The smoke produced during the burning of any fuel contains charged particles. When these charged particles are passed through a charged capacitor, they get attracted towards the plates of the capacitor. Such a device is called an electrostatic precipitator. The chimney described in this project acts as an electrostatic precipitator. When high voltage is applied on the chimney and a wire hanging at its centre, it acts as a precipitator. All the charged particles of smoke stick to the walls of the chimney and so smoke pollution is prevented.

You Require

- *One cardboard tube about one foot long and one inch diameter*
- *Thin aluminium foil 4" wide*
- *One induction coil of high voltage*
- *One matchbox*
- *One cigarette*
- *One switch*

What To Do

- Take one cardboard tube about 12" long and 1" in diameter. Drill a hole of about 1" diameter near one of its ends.

- Take aluminium foil about 4" wide and wrap it around the cardboard tube near the other end.

- Hang one thick metal wire down the centre of the tube such that it does not touch the walls of the tube.

- Connect the foil with the high voltage terminal of an induction coil through a switch and the hanging wire to the other terminal of the induction coil.

- Bring a lit cigarette near the hole of the tube, the smoke will come out from the chimney.

- Now, put on the induction coil in the circuit and again bring the lit cigarette near the hole. In this case, the chimney will not give out any smoke.

Note: *This project can not be done with low voltage induction coil.*

Wire

Cardboard chimney

Induction coil

Cigarette

Battery

connections made

– induction coil

– battery

76

44. Making an automatic letter alarm

The simple circuit described in this project is based on the fact that a bulb glows when there is a letter in the letter box. It is an on/off device operated by the weight of the letters. When the switch is pressed, if the bulb glows, there is a letter in the box otherwise not.

You Require

- *One letter box (16" × 10" × 5")*
- *One electric bell*
- *One bulb of 15 W 220 V with a bulb holder*
- *Plastic-coated wires*
- *Tin plates about 1.5 cm wide*
- *4 screws of one inch size*
- *Sellotape or black tape*

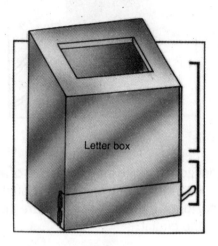

Fig. 1

What To Do

- Get a letter box made of wood by a carpenter having a size of 16" × 10" × 5". The top of the box from where letters are to be dropped in should be at an angle of 45° (Fig. 1). The opening of the box should be in front, as shown in Part 3 of Fig. 1.

- Before fixing the front side of the box, all the electrical connections should be made.

- About 7" from the top inside the box, a tin plate made in half-U shape be fixed with a screw in such a way that the curved side must face you. About one centimeter below, there should be another straight tin plate fixed with a screw. These two tin plates should not block the way of the letter. They should be fixed in such a way that the letter slides down easily. A bell should be connected with the plates. The bell will ring with a little touch of the letter to the plates. As soon as the letter slides down-wards, the bell stops ringing.

- At the bottom, about 1" above of the box, two tin plates 8" each should be screwed

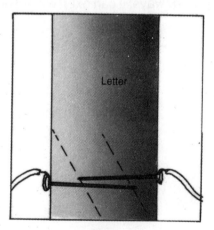

Fig. 2

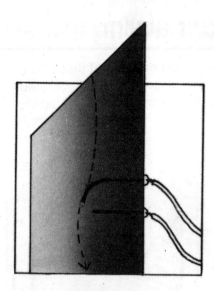

Fig. 3

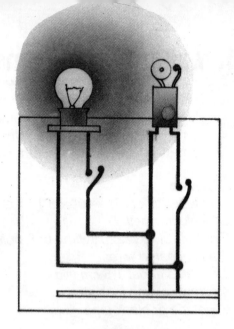

Fig. 4

horizontally with only a little gap of about 3 to 4 mm. These tin plates should also not touch each other as they are the key of the circuit.

- To the horizontal tin plates, two other small tin plates one to one horizontal ones should be fixed, so that if the letter falls sideways, it can exert weight on the

plates for the circuit to be completed.

- A switch should be connected in the bulb circuit so that it does not glow continuously.

- When the wiring is complete and a letter is dropped into the box, the bulb will glow and the bell will ring.

■■

45. Making a Morse code for sending messages

The Morse code telegraph makes use of dots and dashes for different letters and numbers. Using dots and dashes in different combinations the letters of the alphabet and the numbers can be made as shown in Fig. A. The Morse code telegraph brought about a great revolution in communications in the second half of the nineteenth century. People could exchange messages almost instantly over long distances by this telegraph in any kind of weather, day or night.

You Require

- *8 pieces of wood (10 cm × 5 cm × 3 cm)*
- *Steel screws*
- *Nails*
- *A small piece of cloth*
- *Two steel hinges (5 cm × 2 cm)*
- *Two thin saw blades about 8 cm long*
- *Insulated copper wire of 22 to 24 gauge*
- *One 4.5 volt battery*
- *Three dry cells connected in series*

What To Do

- Take a saw blade to make a morse key. Screw the key to one end of the wooden piece but do not tighten the screw yet. Put another screw in the other end of the wooden board so that the key almost touches it (Fig. A).

- Make a coil by winding 150 turns of insulated copper wire on a screw. Screw the coil into place on a piece of wood.

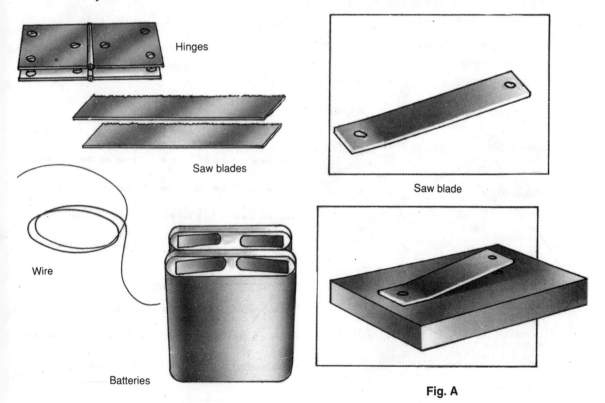

Hinges

Saw blades

Wire

Batteries

Saw blade

Fig. A

79

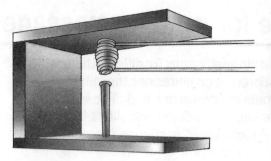

Fig. B

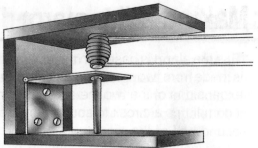

Fig. C

- Make the sounder by mailing the piece of wood, containing the coil to two other pieces of wood as shown in the Fig. B.

- Now, fix the hinge as shown in Fig. C. It should be supported by the nail. Leave a gap of about 2 mm between the hinge and the coil screw.

- Make two keys and two sounders. Connect them to the battery as shown in Fig. D. Now tighten the screws at the fix end of the metal keys.

Test the system by pressing Key 1 so that the free end of the metal blade touches the screw underneath it. This should make Sounder 2 click. Test again with Key 2 and Sounder 1. If everything works in order, you can send any message using the morse code.

When you press a key, the current flowing through the coil wire makes the screw magnetic. It attracts the free arm of the hinge which moves upwards and makes a click. When the key is released, the hinge, falls back.

Morse Code

Different letters and numbers are transmitted by following dots and dashes:

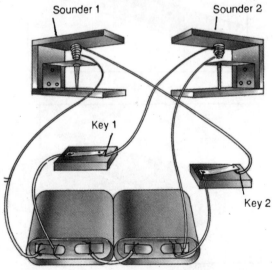

Fig. D

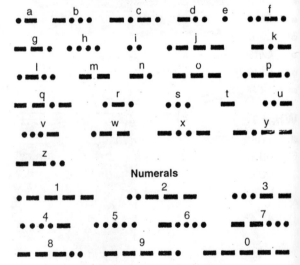

46. Making a fire alarm

This fire alarm is based on the action of a bimetallic strip. The bimetallic strip is made from two metals having unequal expansions. In case of fire, unequal expansions of the two metals in the bimetallic strip cause it to bend so that it completes a circuit to sound a buzzer.

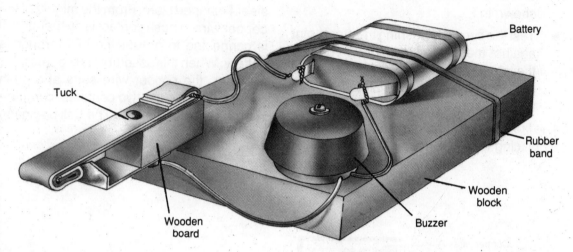

You Require

- *One thin strip of steel*
- *One thin strip of aluminium*
- *Insulated copper wire*
- *One wooden block*
- *One wooden board*
- *One drawing pin or tag*
- *Glue*
- *One small electric buzzer*
- *One battery*
- *Screws*

What To Do

- Take two strips—one of steel and the other of aluminium. Put the steel strip below the aluminium strip. Hold them together to make a good contact. Use a

Fig. A *Bimetallic strip*

pair of pliers to fold the ends over, as shown in Fig. A.

- Make an electrical connection at one end of the bimetallic strip. Then fold this end of the strip over so that the end of a piece of wire is gripped firmly by the aluminium

Drawing pin or tuck

- Take another strip of steel and shape one of its ends to a point and bend it, as shown in Fig. B. Scrape near the other end for a connection to be made. Make a connection at the other end of the copper wire.

- The bimetallic strip and the steel strip must be mounted by tacking them to a

81

block of wood. They must almost touch. Glue the block with the metal strips attached to the wooden board.

- Take a small electric buzzer and a suitable battery for it. Mount the battery with a rubberband and screw the buzzer to the board. Complete the circuit as shown in Fig.

- Test the circuit by pushing the steel point against the bimetallic strip. As soon as the steel point touches the bimetallic strip, the circuit gets completed and the buzzer sounds.

The fire alarm working can be demonstrated by placing it near a heater. The aluminium will expand more than the steel in the bimetallic strip. This will make the strip bend and touch the steel point. With the circuit thus completed, the buzzer will automatically sound.

Note: You can make another model of a fire alarm by using a copper wire mounted on a steel T-shaped plate. From the middle of the copper wire, a push contact is suspended. It is connected to an electric bell through a battery. When this assembly is brought near a heater, the copper wire sags and push contact drops to the two contacts, completing the circuit of the bell. The bell begins to ring.

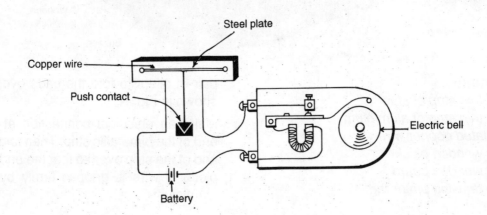

47. Switching on a table lamp by a match stick

This project is based on the phenomenon of photoelectric effect. When a match stick is lit and its light falls on the photo detector or LDR, the electric circuit gets completed and as a result the electric bulb starts giving out light.

You Require
- *6 volts D.C. battery*
- *One table lamp fitted with 6 volt bulb*
- *One LDR or photo detector*
- *One transistor 2N 1177*
- *50 kilo-ohm potentiometer*
- *One switch*
- *One plastic sheet*

What To Do
- Take a plastic sheet and connect all the components as shown in the figure.
- Connect two terminals of the table lamp with the circuit as shown in the figure.

- Cover the photo detector or LDR with a metal cap so that no outside light falls on it.

- When you wish to demonstrate the switching on of the bulb, make the switch on and lit the match stick and bring it near the LDR or photo detector. The bulb will start glowing.

- As long as the light from the bulb of the table lamp is falling on the detector, the circuit will remain on, when you wish to put-off the bulb you have to put-off the main switch of the circuit.

■ ■

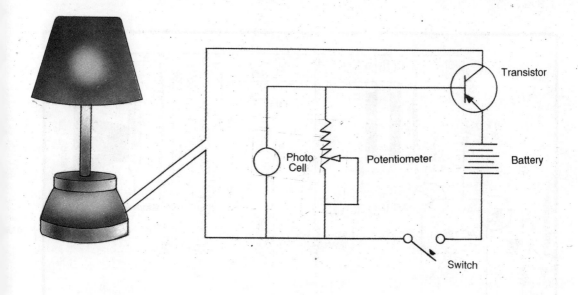

48. Making an electronic burglar alarm

The burglar alarm described in this project is a simple solid state device and can be used to detect burglars and intruders. This is based on a light-sensitive transistor. As long as the phototransistor receives light from a bulb, the circuit does not produce any sound by the loudspeaker. This happens because as long as the light falls on the phototransistor, the base emitter circuit of the transistor AC 128 is short and does not send any current to the relay. But when the light falling on the phototransistor is cut, the conductivity of the transistor is reduced and a current goes to the relay and starts it. The circuit comes into action and the sound is produced by the loudspeaker, indicating that someone has interrupted the light.

You Require
- One plastic box
- One phototransistor AC 126
- One transistor AC 128
- One transistor 25B77
- One relay 6 volt 3-5 mA
- Two capacitors one of 0.5 mF and the other 0.1 mF
- Three resistors — two of 10 and one of 1.9 kΩ

- One output transformer
- One loudspeaker 4Ω
- 5 dry cells, connecting wire
- One bulb of 60 watt

What To Do
- Take one plastic box without a lid to mount the electronic components.
- Take one AC 126 transistor and carefully

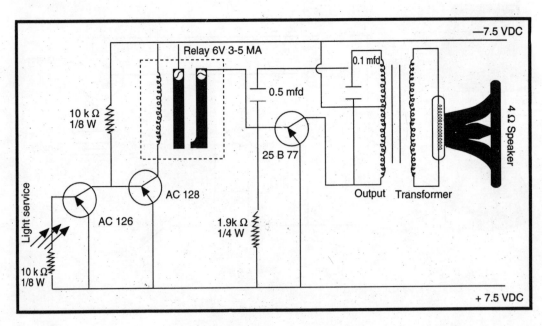

Circuit diagram of burglar alarm

remove its metallic cap. Then wash the inside with carbon tetrachloride till the white paste-like substance is completely removed. Now, cover the transistor with a transparent glass cap to protect it from moisture and damage. In this way, your phototransistor is ready.

- Complete the circuit on the base of the plastic box, as shown in the circuit diagram.
- Connect five dry cells in series to get 7.5 volt. Connect this combination to the circuit.
- Put on the 60 watt light bulb and make its

light fall on the phototransistor. As long as sufficient light is falling on the phototransistor, the circuit will not be activated.

- Now, cut the light by introducing some piece of cardboard between the bulb and the phototransistor, the circuit will become active and you will hear a sound.

Note: You can make a simpler burglar Alarm as follows. This can be mounted at the door and make and break switch can be put under a doormat. As soon as the switch is pressed by the feet of some one, the bell will start ringing. ■■

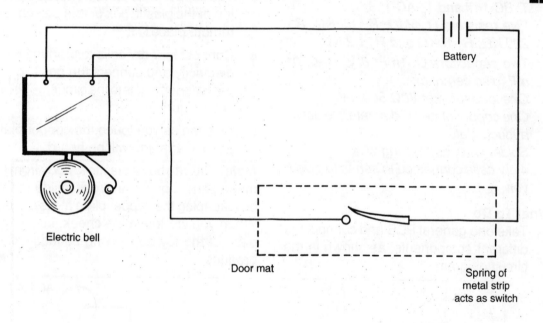

Electric bell

Battery

Door mat

Spring of metal strip acts as switch

49. Making a touch alarm

The touch alarm described in this project is a wonderful electronic device. When you just touch a copper plate mounted at the top of the lunchbox, peculiar sounds are produced.

The circuit comprises two stages. The first stage having T_1 and T_2 transistors forms a trigger circuit. When the copper plate is touched, the transistor T_1 conducts and produces feeble pulses which are amplified by T_2. The second circuit is an audio oscillator made by T_3 and T_4. The signal from the trigger part becomes an audio signal which is converted into sound by the loudspeaker.

You Require

- One plastic lunchbox
- One general PC board
- Four transistors T_1-BC 149; T_2-BC 149; T_3-BC 147 and T_4-AC 128
- Five resistors R_1-5MΩ; R_2-100KΩ; R_3-220 KΩ; R_4-10KΩ and R_5-1.2 KΩ
- Two capacitors C_1-1mF; 6 V; C_2-0. 01 mF (disc ceramic)
- One loudspeaker 80Ω 500 mW
- One copper plate (1" diameter) to act as a touch plate
- Solder wire, connecting wire
- 4 dry cells connected in series to give 6 volt

What To Do

- Take one general PCB and connect the different components, as shown in the circuit diagram.
- Mount the PCB in the plastic lunchbox.
- Mount the loudspeaker at the top of the plastic box.
- The copper plate can be mounted on the top of the plastic box or can be kept at a remote place.
- Connect four dry cells to produce 6 volt electricity and connect the positive and negative terminals to the circuit, as shown in the figure.
- As soon as you touch the copper plate, a peculiar alarm will be heard.

Note: You can buy a plastic toy boy from the market and mount the circuit inside the toy boy, keeping the copper plate at his cheeks. When you will touch the cheek, i.e. copper plate of the toy boy, crying sounds will be produced. ■■

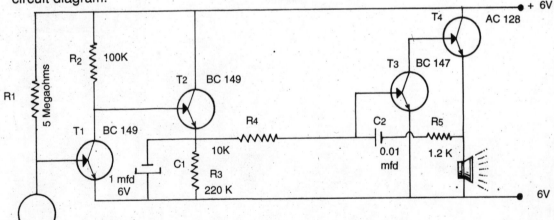

Circuit diagram of touch alarm

50. Making an electronic timer

The electronic timer is a device which controls the on/off time of an electrical appliance. Electronic timers are used in photographic darkrooms, electroplating plants, washing machines, X-ray machines, etc.

The electronic timer described in this project consists of two parts — the power supply circuit and the timer circuit. The power supply part converts 220 volt AC into 15 volt DC. Transformer T_1 converts 220 volt AC into 17 volt AC, which is rectified by the diode bridge D_1, D_2, D_3 and D_4. The output of the bridge circuit is smoothed by the filter circuit consisting of capacitor C_1. A three-pin voltage regulator LM 7812 provides regulated voltage of 12 volt DC.

The timer circuit makes use of IC 555. It has 8 pins. In the circuit, a switch has been connected at Pin No. 2 of the IC and a relay, LED at Pin No. 3. The circuit can be wired on a general purpose PC board. As the switch S_1 closes, the voltage appears at Pin No. 3 of the IC. This output operates the relay and makes the yellow LED off and green LED on. The time of relay is controlled by R_1, R_2 and C_1. The time can be set in this circuit from 1 second to 1 minute.

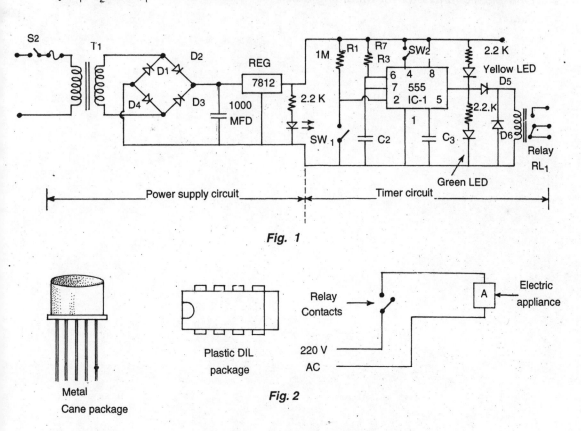

Fig. 1

Fig. 2

Circuit diagram of electronic timer

87

You Require

- *General purpose PCB*
- *Soldering wire, connecting wire*
- *Transformer T_1-primary 220 V Secondary 17 volt*
- *Six diodes D_1, D_2, D_3, D_4, D_5 and D_6-1 N 4001*
- *Three capacitors C_1-1000 mF, 50 V, C_2-100 mF, 50 V, C_3-0.01 mF*
- *Three resistors R_1-1mΩ 1/4 W, R_2-10 KΩ 1/4 W, R_3-Carbon potentiometer 1000 kΩ. One IC 555, one voltage regulator LM 7812-Reg*
- *One relay RL_1-12VDC single contact switch SW_1*
- *One on/off switch SW_2*
- *Three LED — yellow, green, and red*
- *One plastic lunchbox for mounting the circuit*

What To Do

- Arrange the electronic components on a PCB and enclose it in a plastic box. LED's should be mounted on the top of the box.
- Press the switch SW_2, the yellow LED will glow.
- Connect the appliance, say a bulb, and set the time on the potentiometer. Press the switch SW_1, the yellow LED will go off and green will be on. The relay will be energised. The AC supply of 220 volts will be fed to the appliance through this relay.

After the elapse of the set time the relay will get de-energise, yellow LED will be on and green LED will go off.

■■

51. Making a portable metal detector

This device can be used to detect or locate hidden metal objects buried in the ground or in the body.

The circuit consists of two oscillators, both working at about 455 kHz. The first oscillator uses an IFT coil of a transistor radio as the tank circuit. The other oscillator uses a search coil and a variable capacitor as the tank circuit. Both these oscillators are coupled together by a capacitor. When a metal object comes near the search coil, the frequency of the oscillator gets changed. This causes a beat note to be produced, which is detected by the diode and goes to the one transistor amplifier where an audio tone is heard from the speaker.

The detector should be calibrated before use. With the instrument 'on', place the search coil near the metal object. Then adjust VC_1 till a tone is heard. Now, move the metal object away from the search coil, the tone should disappear. It should reappear and increase in loudness as the metal object is brought nearer and nearer.

The search coil has about 20 turns centre tapped, (30 gauge, enamelled copper wire) wound on a 6×6 inch frame. It should be wrapped with insulation tape and fixed with Araldite to a piece of plyboard. A handle should be fixed at the centre of the plyboard and the rest of the electronics mounted in a box and clamped to the handle, as shown in the figure.

You Require

- One plastic lunchbox
- Rigid plastic tube
- One plyboard $(7" \times 7")$
- One wooden frame $(6" \times 6")$ to make search coil
- Eight capacitors C_1-0.47 mF, C_2-22pF, C_3-390 pF, C_4-22pF, C_5-part of the IFT, C_6-0.01 mF, C_7-5 mF, 10 VDC, C_8-100 mF 10 V DC
- 2 Joule gang capacitor VC_1
- Search coil-D_1 to be wound on the wooden frame
- Loudspeaker 15 Ω(Ls)
- Battery B_1-9V
- Switch on/off S_1
- Five resistors R_1-1mΩ 1/4 W, R_2-2 kΩ 1/4W, R_3-1mΩ 1/4 W, R_4-10kΩ 1/4 W, R_5-27kΩ 1/4 W
- Three transistors TR_1, TR_2 and TR_3-BC 109C
- Diode D_1-0A91 wire

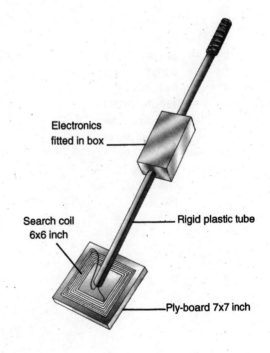

Electronics fitted in box

Search coil 6x6 inch

Rigid plastic tube

Ply-board 7x7 inch

What To Do

- Arrange the circuit in a plastic lunchbox as shown in the circuit diagram.

- Take one wooden frame (6"× 6") and make a search coil of about 20 turns centre tapped (30 gauge enamelled copper wire). Wrap it with insulation tape

and fix it with Fevicol on a piece of plyboard (7"× 7"), as shown in the figure.

- Fix a handle of plastic tube at the centre of the plyboard and fix the plastic box containing the electronics with this tube.

- Now, if you bring the search coil near some metal object, you will hear a sound. ■ ■

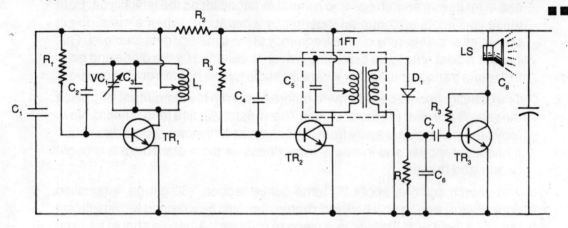

Circuit diagram of portable metal detector

Parts List

C_1	—	0.47 µF	LS	—	Loud Speaker
C_2	—	22 pf			15Ω, Min.
C3	—	390 pf	B1	—	Battery, 9V
C4	—	22 pf	S1	—	ON/OFF Switch
C5	—	part of IFT	R_1	—	1MΩ, 1/4W
C6	—	0.01µF	R_2	—	2KΩ, 1/4W
C7	—	5 µF, 10V DC	R_3	—	1MΩ, 1/4W
C8	—	100 µF, IOV DC	R_4	—	10KΩ, 1/4W
VC1	—	2J Gang Condenser	R_5	—	27KΩ, 1/4W
		(only one part used)	TR1	—	BC 109C
L1	—	Search Coil	TR2	—	BC 109C
		(See Text)	TR3	—	BC 109C
IFT	—	Ordinary Transistor	D1	—	OA91
		IFT			

52. Making a telephone recording interface

All telephone calls can be recorded automatically on a cassette recorder using this interface.

Telephone lines A and B are connected to a voltage sensing circuit through a bridge circuit, to allow for either polarity of line voltage. When the telephone is not in use, the line voltage exceeds about 40V and TR_1 high voltage transistor is turned on. The Arlington pair connected transistors TR_2 and TR_3 are turned off, the relay is not energized, and thus the contacts of the relay are open. During a call, the line voltage falls below 40V, so that TR_1 is turned off. TR_2 and TR_3 are turned on, the relay is energized, and its contacts are closed. These contacts connect to a jack plug fitted to a recorder remote control input.

Speech from the telephone is fed through a 100 nF blocking capacitor and stepdown audio transformer to a jack plug on the recorder that may be operated when the telephone line is not in use without unplugging the circuit, e.g. for rewinding and playback. The circuit could be adopted to switch the mains power to the recorder. When the telephone is not in use, current drawn from the 9V battery is very small.

A plug may be fitted to the output of the recorder for replaying recording over the telephone line.

The voltage sewing circuit draws very little current from the line, when the telephone is not in use.

You Require

- *One tape recorder*
- *One plastic lunchbox for arranging the circuit*
- *Three resistors R_1-1MΩ 1/2 W, R_2-15KΩ 1/2 W, R_3-1MΩ 1/2 W*
- *One capacitor C_1-100 nF ceramic*
- *Five diodes D_1, D_2, D_3, D_4-IN 4004 and D_5-IN 4001*
- *Three transistors TR_1-2N3440, TR_2 and TR_3-BC-107*
- *One 9 volt relay with a coil resistance 2500 Ω*
- *One on/off switch S_1*
- *One audio stepdown transformer T_1*
- *One jack plug for recorder*

What To Do

- Take one plastic lunchbox and arrange the circuit components on its bottom. You can also use a printed circuit board.
- Connect A and B points of the circuit to the telephone line.
- Connect the output of the audio stepdown transformer to a jack plug on the recorder microphone input.
- The message coming through the telephone will be recorded on the tape in the recorder.

■■

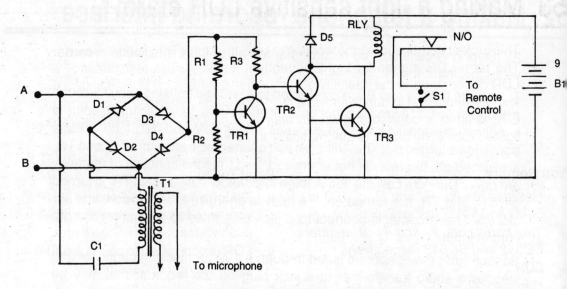

Circuit diagram of telephone recording

Parts List

R_1	—	1MΩ, 1/2W	D_4 —	1N4004
R_2	—	15KΩ, 1/2W	D_5 —	1N4001
R_3	—	1MΩ, 1/2W	TR_1 —	2N3440
C_1	—	100nf ceramic	TR_2 —	BC 107
D_1	—	1N4004	TR_3 —	BC 107
D_2	—	1N4004	RLY —	9V Relay
D_3	—	1N4004		Coil resistance 2500 Ω
			S_1 —	ON/OFF switch

53. Making a light sensitive LDR alarm

This project demonstrates the working of LDR through a light sensitive alarm. The term LDR stands for Light Dependent Resistor. The resistance of a LDR depends upon the intensity of light. When light falls on LDR, it shows a low resistance but in dark it shows a high resistance. When the box in which LDR is fitted is opened, it gives an alarm. If any body opens the box in a lighted room, the alarm is automatically turned on.

You Require

- Resistances R_1, R_2 and R_3 of 100 kΩ, 56kΩ and 10kΩ respectively.
- Two transistors T_1 and T_2 of numbers BC558 and BC548 respectively.
- LDR
- Capacitor C_1 – 0.04 μF
- 3vDC (Two cells)
- Speaker – 2.5″, 8Ω .

What To Do

- Two transistors NPN and PNP are connected as a complementary pair with positive feedback R-C oscillator.
- LDR is connected to points A and B and given voltage to base of NPN.
- Output of PNP is connected to speaker.
- When the light falls on LDR, alarm sets off.

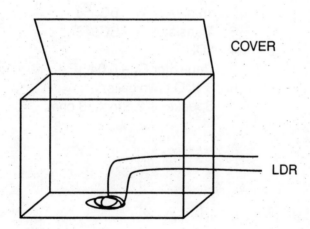

COVER

LDR

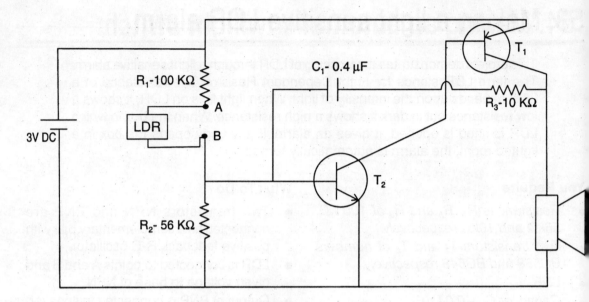

Components

1. Resistances
   R₁ – 100 kΩ
   R₂ – 56 kΩ
   R₃ – 10 kΩ
2. Transistor T₁ – BC558
3. Transistor T₂ – BC548
4. LDR
5. Capacitor C₁ – 0.04 µF
6. 3 vDC (Two cells)
7. Speaker – 2.5 inch, 8 ohm.

54. Making an automatic twilight switch

This switch can be used for controlling the automatic operation of street lights, as porch light. This circuit switches on the lights at dusk and switches them off at dawn automatically. It can also be used on tall buildings and T.V. towers, etc. where a red light is used to warn the approaching aircraft.

Light dependent resistor (LDR) is used to sense the ambient light level to switch the triac on or off. When light falls on the LDR, its resistance becomes so low that the voltage drop across it is negligible and the voltage at Pin 2 of the IC is little less than that of the supply.

The IC functions as a schmidt trigger so that the lamp does not flicker during the on/off operations. The supply to the IC is made fixed (5.IV) by the zener diode, so that the trigger and threshold levels of IC are fixed at $1/3 \times 5.1 = 1.7$ V and $2/3 \times 5.1 = 3.4$ V respectively. With ambient light falling on the LDR, the voltage at Pin 2 is more than the threshold voltage (3.4 V) and thus the output of the IC is low, so that the lamp is turned off. As the dusk approaches, the intensity of light falling on the LDR becomes low. This reduces the voltage at Pin 2 of the IC. When this voltage gets reduced below the trigger level (I.7V), the IC output goes high. This provides the trigger age current for the triac. The triac starts conducting and thus, switches on the lamp load. The lamp once lighted remains 'on' due to the hysteresis provided by the schmidt trigger action, till the intensity of the light falling on the LDR increases sufficiently so as to increase the voltage at Pin 2 of the IC above 3.4V.

D_1, D_2, C_1, R_1 and Z_1 provide the necessary DC low voltage. Heat sink should be provided for the triac. The maximum load for type STO 44 triac should not exceed 500 watt. The potentiometer VR_1 is for sensitivity control.

This should be adjusted for optimum results. The LDR should be mounted in such a way that is does not receive the sunlight directly but only natural light. No other light should interfere with the LDR's operation. It should also be covered with transparent glass to protect it from dust and rain.

You Require

- One plastic box
- One PCB
- Three resistors — R_1-15 Ω1 watt, R_2-1kΩ $1/4$ watt, R_3-22Ω $1/4$ watt
- Two capacitors C_1-500μF, 16V, C_2-0.01 mF, 16V
- One light dependent resistor (LDR)
- One 10 kΩ potentiometer (VR_1), two diodes D_1-IN 4001 and D_2-IN 4002

- One zener diode Z_1-5.IV, IW
- One IC-LM 555
- One transformer T_1-primary 220 volt
- Secondary 6-0-6 volt 250 mA
- Triac-400V, 4A (STO 44 or equivalent)
- One one-off switch (S_1)
- One neon indicator lamp(N_1)
- Connecting wires
- Soldering iron
- Soldering wire

What To Do

- Take a printed circuit board and connect the electronic components with each other as shown in the circuit diagram.

- Mount the triac with a heat sink.

- The load can be chosen as 5 electric bulbs of 220 volt and 60 watt each. They should be connected in parallel.

- The printed circuit board can be mounted in a small plastic lunchbox.

- The sensitivity of the circuit can be adjusted by the variable potentiometer. ■■

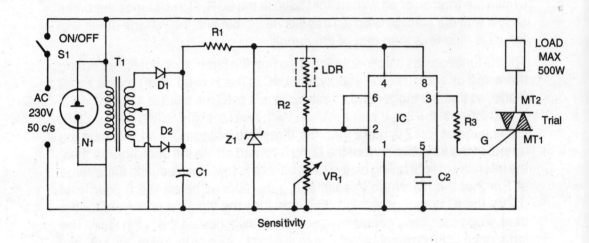

Sensitivity

Parts List

R$_1$	—	15Ω, 1/4W	VR$_1$	—	10KΩ, potentiometer
R$_2$	—	1kΩ, 1/4W	D$_1$	—	IN4001
R$_3$	—	220Ω, 1/4W	D$_2$	—	IN4002
C$_1$	—	500µF, 16V	Z$_1$	—	5.1V, 1W, Zener diode
C$_2$	—	0.01µ, (Disc, ceramic)	IC	—	LM 555
LDR	—	Light dependent Resistor	Triac	—	400V, 4A, Type STO44 or equivalent
			T$_1$	—	6-O-6V, 250 mA Sec Transformer

55. Making a radio set (using a semiconductor)

The radio waves coming from a radio station are picked up by an antenna. The waves cause varying electric currents to flow down the wire. These currents pass through the radio set and into the ground.

Inside the radio set is a tuned circuit consisting of a capacitor and a coil. The tuned circuit is used to select the required signals from the mixture picked up by the antenna. A diode connected between the tuned circuit and earphone detects and demodulates the signal. The earphone converts it into sound and we hear the programme being relayed by the radio station.

You Require

- *Enamelled copper wire*
- *Plastic tube*
- *Tape*
- *One capacitor of 30 pF*
- *One semiconductor diode*
- *One earphone*
- *One wooden board*
- *Thumb tacks*
- *One staple*
- *One long metal rod*

What To Do

- Take enamelled copper wire of 22 to 24 gauge and wind a coil of about 100 turns on a plastic tube (1" to 2" diameter and 4" long). To make the coil, start by attaching the end of the wire to one end of the tube

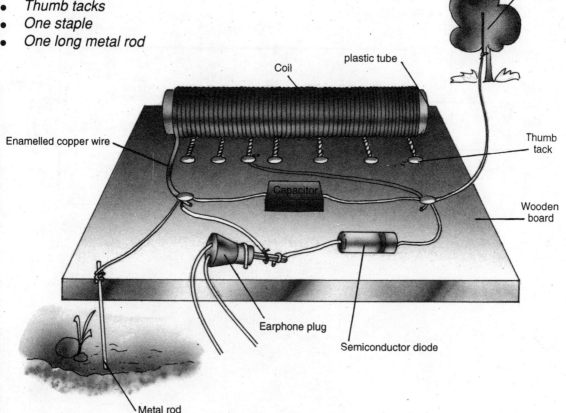

Antenna

plastic tube

Coil

Enamelled copper wire

Thumb tack

Capacitor

Wooden board

Earphone plug

Semiconductor diode

Metal rod

with tape or make two small holes in the tube and insert the wire through them. Then wind the wire neatly around the tube so that the turns lie side by side. At intervals, twist the wire to form a 'tail' protruding from the coil, and then continue winding. Make about six of these tails which should be closer together near one end of the coil.

- Purchase a 300 *pF* fixed capacitor and a semiconductor diode. For being able to listen to the received radio signals, you have to buy an earphone also.

- Fix the components on a wooden board and make the connections as shown in the figure. They can be fixed with thumb tacks. A staple is required to fix the earphone plug.

- Scrape the enamel from the ends of the copper wire and tails before fixing them in place.

- Run a wire out of house and attach it to a convenient high point. This will act as the antenna.

- Take one metal rod and put it about 2 feet into the earth. Attach it to a wire.

- To receive signals, touch the free wire onto the coil connections, one after another. By doing so, you will be able to tune in to a powerful radio station and listen to the programme being broadcast.

Circuit Diagram of the Radio Set

The circuit diagram of this diode radio set is given below:

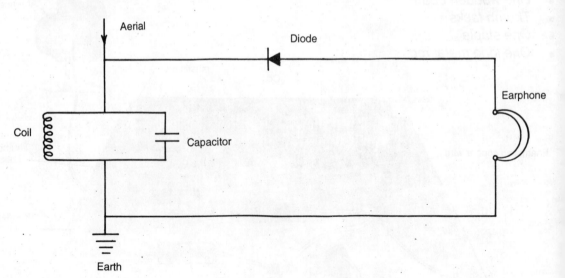

56. Making a radio set (using a variable capacitor)

You Require

- *Enamelled copper wire (22 to 24 gauge)*
- *Plastic or cardboard tube (1" to 2" diameter and 4" long)*
- *One semiconductor diode*
- *One variable capacitor*
- *One headphone*
- *One wooden board (6" × 4")*
- *A piece of sandpaper*

What To Do

- Wind about 20 turns of enamelled copper wire on the plastic or cardboard tube. Take the two free ends of the coil and remove the enamel by rubbing with a piece of sandpaper.

- Make the connections of the different components on a wooden board, as shown in the figure.

- Use some length of copper wire as aerial.

- Put the headphone on your ears and rotate the knob of the capacitor very slowly. You will catch some radio station and be able to listen to the programme.

■■

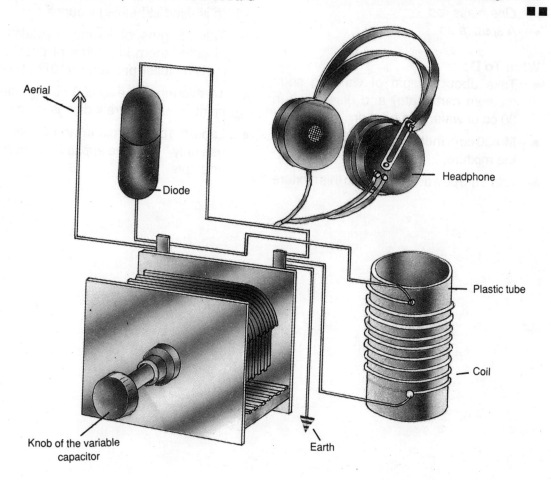

Aerial — Diode — Headphone — Plastic tube — Coil — Knob of the variable capacitor — Earth

57. Making iodoform

Iodoform is a derivative of iodine. It is prepared by using iodine, methylated spirit and washing soda. It is a strong disinfectant and is used as an antiseptic. Iodoform is used to kill germs in cuts and scratches. It saved thousands of wounded soldiers during the first World War. You can make iodoform as follows.

You Require
- *Iodine crystals—3 gm*
- *Methylated spirit—30 cc*
- *Washing soda—20 gm*
- *Water—30 cc*
- *A beaker of 250 ml*
- *One glass rod*
- *A spirit lamp*

What To Do
- Take about 20 gm of washing soda (sodium carbonate) and dissolve it in 30 cc of water in a beaker.

- Mix 30 cc of methylated spirit to it and stir the mixture.

- Add 3 gm powder of iodine to this mixture and warm the beaker on the flame of a spirit lamp. Do not heat it. Go on stirring. A yellow shining solid will separate. Let the mixture cool. Dry the yellow solid on a blotting paper. This is used as iodoform.

Note: You can also make tincture iodine which is used in healing wounds.

- Take 5 gms of iodine crystals and dissolve them in 5 gms of potassium iodide solution prepared in 10 cc of water.

- To this mixture add 250 cc of rectified spirit. Your tincture iodine is ready.

- Label the bottle "Poison" and "for external use only." I' can be applied on swellings and cuts.

██

58. Detecting adulteration in *ghee*

Vanaspati ghee is manufactured by the hydrogenation of vegetable oils. In the process of hydrogenation, nickel is used as a catalyst and hydrogen is passed. By this process, unsaturated oils become saturated. To confirm the adulteration of *vanaspati ghee* in pure *ghee*, we conduct the test for nickel particles, which are always present in traces in *vanaspati ghee*.

You Require

- *One test tube*
- *Hydrochloric acid*
- *Pure ghee mixed with vanaspati ghee*
- *2% solution of furfurol in alcohol*
- *Spirit lamp*

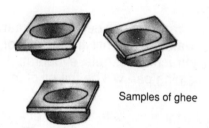

Samples of ghee

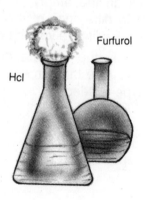

Hcl   Furfurol

- Pour an equal quantity of hydrochloric acid in the test tube and shake the mixture well for about one minute.

- Now, add a few drops of 2% alcoholic solution of furfurol in the test tube and shake it for about 2 to 3 minutes.

- Allow the mixture to settle down for about 2 minutes. If you see pink colour in the test tube, it indicates the presence of *vanaspati ghee*.

■ ■

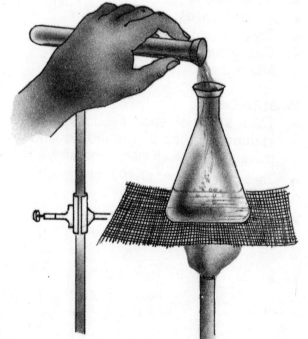

What To Do

- Take about 5 cc of pure *ghee* mixed with a little quantity of *vanaspati ghee* in a test tube. Heat it on the flame of a spirit lamp.

59. Extracting fat from oilseeds

Fats are organic and non-polar in nature, so they get dissolved in a solvent like carbon tetrachloride. Adding carbon tetrachloride to the crushed oilseeds, we can dissolve the fats present in the seeds. On filtering and evaporating carbon tetrachloride from the filterate, the fats can be separated. The percentage of fats present in the seeds can be determined by weighing the fat and crushed seeds and taking their ratio, i.e.

Fat percentage = Weight of the fat × 100 / Weight of the crushed seeds

You Require

- *5 test tubes*
- *5 beakers of 100 ml*
- *Pipette*
- *Physical balance*
- *Water bath*
- *Spirit lamp*
- *Funnel*
- *Filter paper*
- *Carbon tetrachloride*
- *Walnut seeds*
- *Almonds seeds*
- *Dry coconut*
- *Groundnut*
- *Mustard seeds*

What To Do

- Take 5 gm of each seed (walnut, almonds, coconut, groundnut and mustard) and crush them separately.
- Take five beakers of 100 ml capacity and pour 20 ml of carbon tetrachloride in each. Add 5 gm of crushed seeds in each beaker separately.
- After about 10 minutes, filter the mixture of each beaker and throw away the residues.
- Take the filtrate of each beaker and heat it on a water bath until the whole of carbon tetrachloride gets evaporated. The remaining part will be the oil of the seeds.
- Weigh the fat along with the beaker and empty the fat into the test tubes. Then weigh the empty beakers separately. The difference of the weights in each case will give you the quantity of fats.
- Apply the formula given above and calculate the percentage of fat in each type of seeds.

■■

60. Finding the composition of water by electrolysis

When electric current from a battery is passed through water, it splits water into its constituents, i.e. hydrogen and oxygen. This is known as the electrolysis of water. The two gases can be collected in two test tubes and the composition of water can be studied.

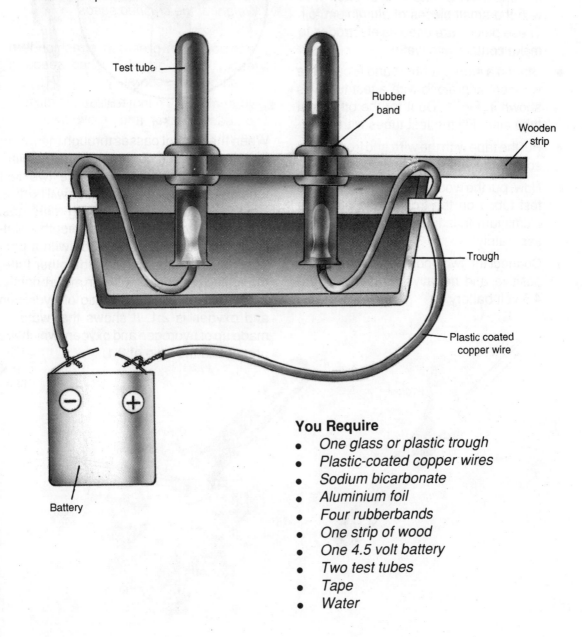

You Require
- One glass or plastic trough
- Plastic-coated copper wires
- Sodium bicarbonate
- Aluminium foil
- Four rubberbands
- One strip of wood
- One 4.5 volt battery
- Two test tubes
- Tape
- Water

What To Do

- Take a glass or plastic trough and put some water in it. To make the water conducting, dissolve one tea spoon of sodium bicarbonate or a few drops of sulphuric acid.

- Take two pieces of plastic-coated copper wire and remove the plastic coating from the ends. Wrap one end of each wire with the small pieces of aluminium foil. These pieces are used as electrodes to make contact with water.

- Stretch a strong rubberband around the wooden strip along with a test tube, as shown in Fig. 1. Do it for the other test tube also. Fill the test tubes with water.

- Put the tape with the wire and trough end so that the wires remain in position.

- Now, put the wooden strip along with the test tubes on the trough and push the aluminium foils into the two test tubes separately.

- Connect the two ends of two wires to the positive and negative terminals of the 4.5 volt battery.

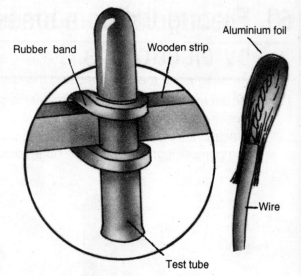

Fig. 1

When the current passes through the water, bubbles of gases come out from the electrodes and gases collect in the two test tubes. The ratio of the two gases will be 1:2. Now, take one test tube having more gas. Bring a lighted matchstick to its mouth, pointing it downward. It will explode with a pop. This gas is hydrogen. Over the other tube, the lighted matchstick will burn more brightly. This gas is oxygen. The ratio of hydrogen and oxygen is 2:1. It shows that water is made up of hydrogen and oxygen, which are present in the ratio of 2:1.

■■

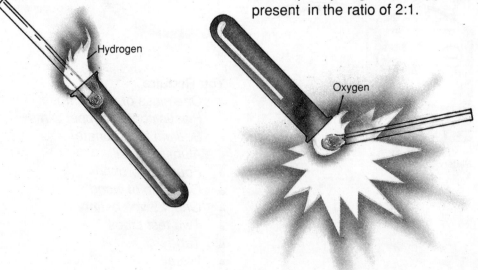

61. Electroplating a brass key with copper

Electroplating is a process by which a thin coating of one metal is applied to another base metal by passing the current. This process improves the look of the base metal and provides it protection against weather. This process is based on electrolysis.

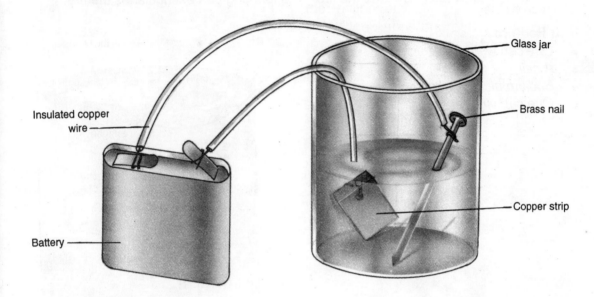

Glass jar

Brass nail

Insulated copper wire

Battery

Copper strip

You Require

- *Copper sulphate crystals*
- *One copper strip (about 3" × 2")*
- *One brass key or brass nail*
- *One battery 4.5 volt*
- *Insulated copper wire*
- *One glass or plastic jar or beaker*

What To Do

- Take a glass or plastic jar. Fill it three-fourth with warm water. Dissolve some copper sulphate crystals in it so that a blue-coloured solution is obtained.

- Take one copper strip, make a small hole in it and connect it to the positive terminal of the battery with an enamelled copper wire.

- Clean the brass door key or nail and connect it to the negative terminal of the battery with the copper wire.

- Dip both the items in the copper sulphate solution in the jar as shown in the figure.

- After a few minutes, you will see that a fine layer of copper gets deposited on the door key. When the key is copper plated completely, remove it and wash it with water.

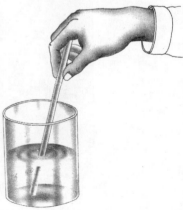

■■

62. Making a fire extinguisher

A soda-acid extinguisher makes use of sodium bicarbonate and acid. The two are normally not allowed to mix. Only when we wish to extinguish a fire, sodium bicarbonate and acid are mixed. This mixture produces foaming carbon dioxide. It is let out through the fire extinguisher in the form of a jet. The jet is directed towards the fire. This foaming gas extinguishes the fire.

You Require

- *One can with a tight-fitting lid*
- *A nail*
- *Sodium bicarbonate powder*
- *One cup vinegar*

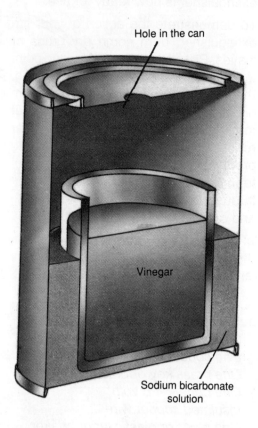

Hole in the can

Vinegar

Sodium bicarbonate solution

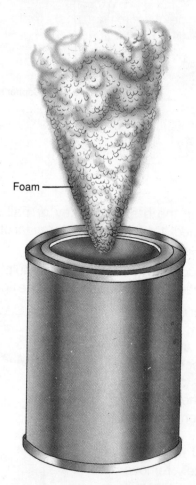

Foam

Fire extinguisher

What To Do

- Take a can with a tight-fitting lid and make a hole in the lid about 3 mm in diameter by placing it on a block of wood and hammering a nail through it.

- Make the saturated solution of sodium bicarbonate in water and pour it in the can.

- Take some vinegar in a small cup and put it in the can. Do not allow the vinegar and sodium bicarbonate solution to mix at this stage. If you find that the solution is more in the can, pour out the excess quantity.

- Place the lid carefully. The fire extinguisher is now ready for use.

- To demonstrate the action of your fire extinguisher, burn some dry grass or leaves. Pick your extinguisher and point the hole towards the fire. By doing so, the two liquids, i.e. the sodium bicarbonate solution and vinegar get mixed. The acetic acid in the vinegar reacts with sodium bicarbonate and produces carbon dioxide. The jet of foaming carbon dioxide comes out through the hole in the can lid (Fig. 1) and extinguishes the fire.

Note: Never point the fire extinguisher toward yourself or anyone else, otherwise a powerful jet of foam will spoil your clothes.

This type of fire extinguishers are used to extinguish ordinary fires. This cannot be used to extinguish electrical fires. In fact water based fire extinguishers should not be used to extinguish electrical fires because this would result in electrocution of the people involved. A fire extinguisher based on the use of carbontetrachloride is used in case of electrical fires. The carbontetrachloride liquid vaporises on coming in contact with the burning material. It is advisable to cover mouth and nose with wet cloth to protect oneself from the effect of carbontetrachloride vapours.

The other type of fire extinguishers that are used for oil and electrical fires are carbon dioxide extinguisher, dry chemical extinguisher and the vaporising liquid extinguisher.

Fire extinguishers are provided by law in public buildings, factories and schools. Most of the big cities have fire brigades for fire prevention and control.

■■

63. Making a model active volcano

In an active volcano, white clouds of steam, ash and lava rise into the sky. You can build a model volcano with vinegar and baking soda. Though there is no fire in this volcano, you will see that lava comes out from the top of the volcano. It will give the appearance of a realistic volcanic cone.

You Require
- *Clay*
- *Water*
- *Plastic tumbler*
- *Baking soda*
- *Detergent*
- *Red colour*
- *Vinegar*

What To Do
- Build a model of volcano out of clay as shown in the figure. Leave some space at the top of this volcano to fit the plastic tumbler in it. Put it in the sun to dry.

- When you want to show its activity, put three or four spoonfuls of baking soda in the tumbler.

produces carbon dioxide gas, which gives the appearance of eruption of a volcano. The red colour creates an effect of volcanic magma and the detergent creates more bubbles.

Underwater Volcano

Sometimes Volcanoes erupt in the ocean. You can make an underwater volcano as follows:

- Take a large glass jar, a small bottle, string and red food colouring. Fill the jar with cold water.

- Fill the small bottle with hot water. Tie a string around its neck and add some red food colouring.

- Holding one end of the string, lower the small bottle carefully into the glass jar and watch as "Lava" rises from the small bottle. This is your mini underwater volcano.

- Mix together half cup of water, one-fourth cup of detergent, one-fourth cup of vinegar and some red colour in another tumbler.

- Pour this mixture into the plastic tumbler in which baking soda was put.

- You will see smoke and bubbles coming out from the cone of the volcano in the same way as they do from a real volcano.

In fact, mixture of baking soda and vinegar

64. An alternative model of a volcano

The chemical volcano is based on the endothermic decomposition of ammonium dichromate. It portrays a volcanic eruption clearly.

You Require

- *One large asbestos wire gauge*
- *Ordinary clay*
- *Wooden stick*
- *Ammonium dichromate 250 gm*
- *Magnesium powder 65 gm*
- *Matchbox*

What To Do

- Take an asbestos wire gauge of 12" × 12" size. (It is available in your chemistry laboratory.)

- Build a volcanic cone with ordinary wet clay about 8" in diameter and 12" in height. Make a 2" deep cavity in the cone with the help of a wooden stick when the clay is wet. Dry it in the sun for two-three days.

- Mix 65 gm of magnesium powder with 250 gm of ammonium dichromate powder.

- Put the asbestos gauge with the volcanic cone on a tripod stand and heat its central part with the flames of a Bunsen burner, as shown in the figure.

- Pour some mixture into the cone of the volcano. The smoke with sparks due to the formation of nitrogen will come out and give the appearance of a volcanic eruption.

■■

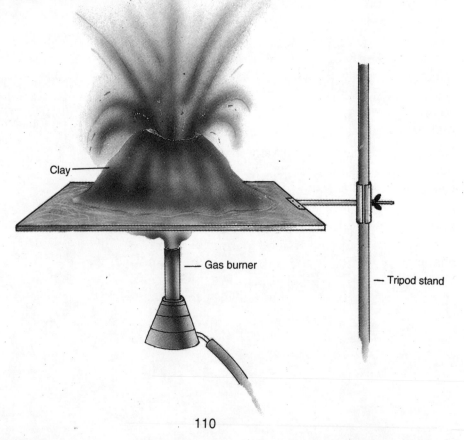

Clay

Gas burner

Tripod stand

65. Demonstrating the destructive distillation of wood

The process of converting wood into its chemical components is called the destructive distillation of wood. When wood is heated, it emits a mixture of gases which burns with a smoky flame. In this project, you can demonstrate the destructive distillation of wood by heating the wood shavings and burning the gases evolved in the process.

You Require

- *One small can with a push-fit lid*
- *One thin nail*
- *Wood shavings or sawdust*
- *Kitchen stove*

What To Do

- Take a small can with a push-fit lid. Make a hole in the lid about 3 mm in diameter by placing the lid on a block of wood and hammering a nail through it.

- Put some wood shavings or sawdust in the can and close the lid tightly.

- Now put the can on a kitchen stove or kitchen gas and heat it gently. After a few minutes, a steady stream of wood gas will be seen coming from the hole in the lid. This gas is a mixture of hydrogen, methane, and other gases.

- Now, light the gas with a match. It will burn with a smoky flame.

- When no more gas comes out, turn off the heat and allow the can to cool. Open the lid and observe what is left in the can. The black solid material is nothing but charcoal.

■■

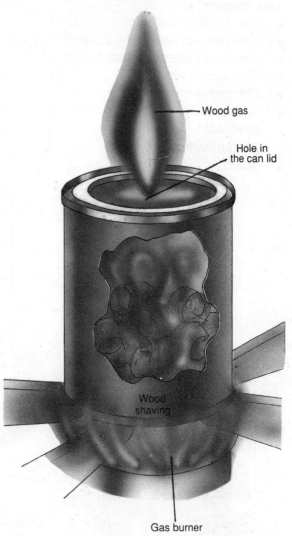

Wood gas

Hole in the can lid

Wood shaving

Gas burner

111

66. Making a chemical photoelectric cell

A photoelectric cell is a device which produces electric current when exposed to light of certain wavelength. Photocells are used as light switches, light detectors and light meters. They are also used in TV cameras, burglar alarms and many other devices. You can make a simple liquid photocell using a copper plate and lead nitrate solution as follows.

You Require

- One beaker of 250 ml
- One copper plate (10 × 2.5 cm)
- One lead plate of the same size
- Connecting wire
- One galvanometer
- Bunsen burner
- Nitric acid
- Lead nitrate
- Torch or Table Lamp

What To Do

- Take one copper strip about 10 cm long and 2.5 cm wide. Bring it near the flame of a Bunsen burner so that by heating, it gets covered with black copper oxide. Cool it and put it in nitric acid until a red layer of cuprous oxide is formed on it. In fact, this layer is sensitive to light and emits photoelectrons.

- Take one beaker of about 250 ml and make a solution of lead nitrate in water.

- Immerse the copper plate and one lead plate in this beaker. Connect the two plates to a galvanometer with the help of a connecting wire.

- Now, throw strong torch light onto the cuprous oxide plate. As soon as the light falls on the plate, the galvanometer will show a deflection, indicating that the electric current has been generated by the falling light rays. If you switch off the light, the deflection in the galvanometer will become zero. In case torch light is not sufficient, you can use a table lamp with a 60 watt bulb.

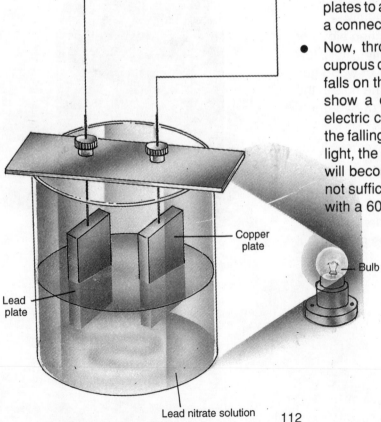

Galvanometer

Copper plate

Lead plate

Bulb

Lead nitrate solution

112

67. Making a chemical garden

A chemical garden is based on the process of crystal growth in the solution of water glass. It is a garden with nonliving plants but that is very attractive.

You Require

- *One glass jar or aquarium tank*
- *Distilled water*
- *Pieces of copper chloride*
- *Copper sulphate*
- *Copper nitrate*
- *Magnesium sulphate*
- *Aluminium sulphate*
- *Ferrous sulphate*
- *Ferrous chloride*
- *Nickel sulphate*
- *Alum*
- *Water glass*

What To Do

- Mix equal volumes of water and water glass and pour this solution into the glass jar or aquarium tank.

- Drop in the solution a few pieces of chemical salts, such as copper chloride, copper sulphate, copper nitrate, magnesium sulphate, aluminium sulphate, ferrous sulphate, ferrous chloride, alum, nickel sulphate, etc. These pieces should be slightly away from one another.

- Place the jar in an undisturbed place. After a few hours, attractive crystalline growths will appear as the chemicals react with the solution. Within a few days, a beautiful garden-type growth will appear in the jar.

68. Growing crystals of copper sulphate

Crystals of copper sulphate, sodium chloride (common salt), sugar, etc. can be grown by the solution technique. A saturated solution of a substance, say, copper sulphate is prepared in 50 ml of distilled water. Some solution is put in a shallow plate for 2-3 days. After a few days, small crystals are seen in the plate. After selecting one good crystal from this plate as a seed crystal and tying it with a thread, it is suspended in a beaker having a saturated solution of copper sulphate. After a period of six-seven days, a large crystal of copper sulphate is obtained.

You Require

- One beaker 200 ml
- One glass rod
- One plate
- Thread
- One pencil
- Distilled water

What To Do

- Take some copper sulphate powder and put it in about 50 ml of hot water in a beaker, stir until no more copper sulphate will dissolve.

- Pour a little of solution into a saucer and allow it to cool without disturbance. After several hours, you will see some small crystals in the saucer.

- Collect one good crystal and tie a fine thread around it.

- Hang this crystal in the copper sulphate solution in a beaker, as shown in the figure.

- Leave the crystal undisturbed for several days. As the water evaporates from the solution, the seed crystal becomes larger and larger. Remove it and dry it with a blotting paper. Remove the thread by cutting it. Your crystal is ready.

Additional crystal growth projects

By using the same method described above, you can try the growth of alum, copper acetate, sodium nitrate, nickel sulphate and sodium chloride crystals.

114

69. Growing plants without soil

Plants grow in the soil because they get all the required nutrients, minerals and water from the soil. They also get support to stand in the soil. If the needed support, nutrients, minerals and water can be provided to a plant externally, it can grow without the soil. This project demonstrates the growth of a plant in a beaker (without soil) in which all the nutrients essential for a plant's growth are provided.

You Require
- One jar or bottle of 1000 cc
- One small plant
- Powder of calcium nitrate (about one gram)
- Potassium nitrate (0.2 gm)
- Magnesium sulphate (0.2 gm)
- Iron phosphate (100 mg)
- Urea (0.2 gm)
- Water (1000 cc)

What To Do
- Take one bottle of one litre capacity and fill it with one litre of water.

- Mix about 1 gram of calcium nitrate, 0.2 gm of potassium nitrate, 0.2 gm of magnesium sulphate, 100 mg of iron phosphate and 0.2 gm of urea. Dissolve all these salts properly in the bottle of water.

- Put a plant in this mixture. You can give a support to the plant by tying a thread with the neck of the beaker or putting a cardboard disc with a hole.

The plant will grow in this solution just as in the soil because it gets all the essential nutrients required for its growth from the solution.

Plant

Bottle · Culture solution

115

70. Demonstrating the feeding of yeast on sugar

A yeast is a fungus. It is a living organism that feeds on sugar and gives off carbon dioxide and either alcohol or water. Yeasts are used in making bread, beer and many carbonated drinks. In this project, you will grow some yeast and prove that it consumes sugar and produces carbon dioxide.

You Require
- 8 Campa Cola or soda water bottles
- 8 balloons
- Yeasts
- Water
- Sugar solution (made by mixing equal parts of water and sugar)

What To Do
- Fill four of the bottles halfway with the sugar solution. Mark these bottles with an 'S'.

- Fill the other four bottles about halfway with plain water. Mark these bottles with a 'W'. Both water and sugar solutions should be at room temperature.

- Add a teaspoonful of yeast to two of the bottles with the sugar solution. Mark these bottles with an 'SY'.

- Add a teaspoonful of yeast to two of the bottles with plain water. Mark these bottles with a 'WY'.

- Put a balloon on the top of each of the eight bottles. Divide them in two groups as shown in the figure. Now, the two groups will be W, WY, S, SY and W, WY, S, SY.

- Place one of these groups of four bottles in the refrigerator. Place the other group in a warm place such as near a heater or radiator.

Make observations every four hours and note the following points:

- What happens to the liquid in each of the eight bottles?

- What happens to the balloons on each of the eight bottles?

- What is the difference in the yeast in plain water and sugar solutions?

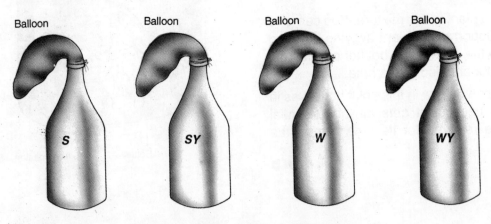

S=Sugar solution   SY=Sugar Solution+Yeast   W=water   WY=Water+Yeast

Bottles with sugar solution

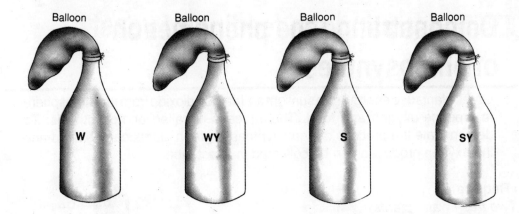

Balloon   Balloon   Balloon   Balloon

W   WY   S   SY

- What is the difference in the yeast growth in the refrigerator and at a warm place?

Try to find out the reasons for these happenings. You will see that in bottles marked with SY and kept in a warm place, the formation and evolution of carbon dioxide is very fast.

Another Way of Demonstrating Yeast in Action

- Take ¹/4 teaspoon of sugar and 2 teaspoons of warm water. Mix them together in a bowl. Sprinkle about 25 gms of active dry yeast.

- Wait for about 20 minutes for the liquid to become frothy. This happens due to the formation of carbon dioxide.

- Now take two cups of flour and one teaspoon of salt. Add water and yeast mixture to the flour. Knead the dough until it feels smooth.

- Now put the dough in a bowl and leave it in a warm place. The dough will swell to the double of its size within one hour.

- Roll the dough to form a 1/4" thick round bread. Bake it in a greased pan for about 20-35 minutes. This whole project demonstrates the yeast in action. Yeast changes the flour starch into sugar then react to form carbon dioxide.

■■

117

71. Demonstrating the phenomenon of photosynthesis

Green plants use energy from sunlight and carbon dioxide from the atmosphere to produce oxygen and food. This process is called photosynthesis. To demonstrate the process of photosynthesis, green plants can be used and the oxygen produced can be collected in a test tube.

You Require

- *Two fresh water plants*
- *One glass funnel*
- *One glass trough*
- *One test tube*
- *Some magnesium strip*

What To Do

- Buy a couple of plants suitable for a fresh-water aquarium from any nursery.

- Take one glass funnel to put over the plants and a test tube to fit over the spout of the funnel.

- Put the plants in a clear glass trough and fill it with water.

- Fill the test tube with water (Fig. 1) and keeping it underwater, slide it sideways over the funnel as shown in Fig. 2.

- Position the funnel and the test tube over the water plants. At this stage, the test tube should still be full with water.

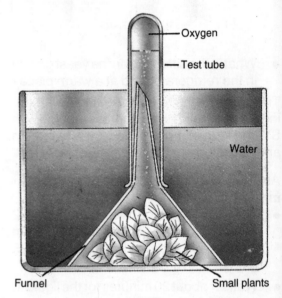

Fig. 2

- Put the trough in a sunny place and observe it for about three hours. The oxygen gas will come out from the plants in the form of tiny bubbles and rise up into the test tube.

- Test this gas by burning a strip of magnesium, which burns more brightly in the test tube, confirming the presence of oxygen.

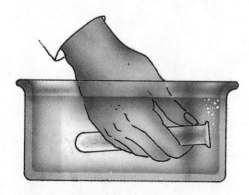

Fig. 1 Fitting the test tube with water

Fig. 3 Glowing matchstick bursts into flame

72. Making an automatic cut off timer

This project is IC based. Time is automatically cut off using IC 555. It works as a monostable timer. When the switch is put on, the timer output is high for a preadjusted time. This output is used to control the electrical appliances with the help of relay driver circuit. When output of IC is low, the connected appliance is off.

You Require

- *Diode D_1 – IN 4001*
- *Capacitor C_1 – 1000 µF/25v*
- *Capacitor C_2 – 1000 µF/25v*
- *Transformer (step down) 220 vAC – 0 – 12v/250 mA*
- *Transistor T_1 – BC 187*
- *Resistance R_1 – 100 kΩ variable*
- *Resistance R_2 – 1 kΩ*
- *Resistance R_3 – 10 kΩ*
- *Relay RL – 6 vDC Relay*
- *Push to switch on sw*
- *Integrated circuit –1 – IC 555*
- *PCB*

What To Do

- Table PCB and mount the stepdown transformer on it.
- Connect one wire from the secondary of the transformer to the capacitor C_1.
- The two wires from the primary should be connected to the relay as shown in the circuit diagram.
- RC component to control the time period of on/off is connected to the pin 6 and 7 of the IC-555.

- Resistance R_2 of 1 kΩ should be connected through transistor T_1 to the relay.
- R_3 resistance – 10 kΩ is connected to sw i.e. push to switch on.
- Make rest of the connections as shown in the figure.

Components

1. Diode D_1 – IN 4001
2. Capacitor C_1 – 1000 µF/25v
3. Capacitor C_2 – 1000 µF/25
4. Q_1 – O – 12v/250 mA
5. Transistor T_1 – BC 187
6. Resistance R_1 – 100 kΩ variable
7. Resistance R_2 – 1 kΩ
8. Resistance R_3 – 10 kΩ
9. Relay – 6vDC Relay
10. SW – Push to switch on
11. IC-I – IC555
12. PCB

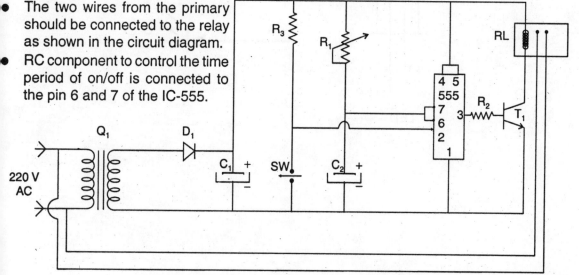

73. Making a direction indicator arrow

This direction indicator arrow is used to indicate the direction of a toilet. Sometimes when the gathering in a party is large and the guests are unable to find the toilet, this arrow/indicator is used to indicate the direction of the toilet. This pointer is very useful for many guests.

You Require

- *Resistances R_1, R_2, R_3, R_4, R_5 – 470Ω*
- *Resistance R_6 – 330Ω*
- *Resistances R_7, R_8, R_9, R_{10}, R_{11} – 1 m2Ω (each)*
- *Capacitors C_1, C_2, C_3, C_4, C_5 – 150 KPF (each)*
- *16 Pin Socket.*
- *D_1 to D_8 – Red LED*
- *PCB*
- *Integrated circuits N_1 to N_6 – IC 1 – 4049*
- *6 ... 12 v Battery or CMOs Inverters*

What To Do

- The circuit is based on a single IC 4049 which contains six CMOS invertors. Each of these has a RC network at its output to provide an appropriate delay before enabling the next invertor.
- The outputs of the chip are capable of driving one LED directly. Delay time of LED is controlled by changing the value of the capacitor.
- Solder all the components on the PCB as per the circuit diagram.
- Test the circuit diagram as a direction indicator.
- The circuit of LED is given as follows:

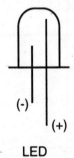

LED

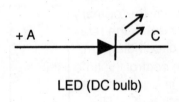

LED (DC bulb)

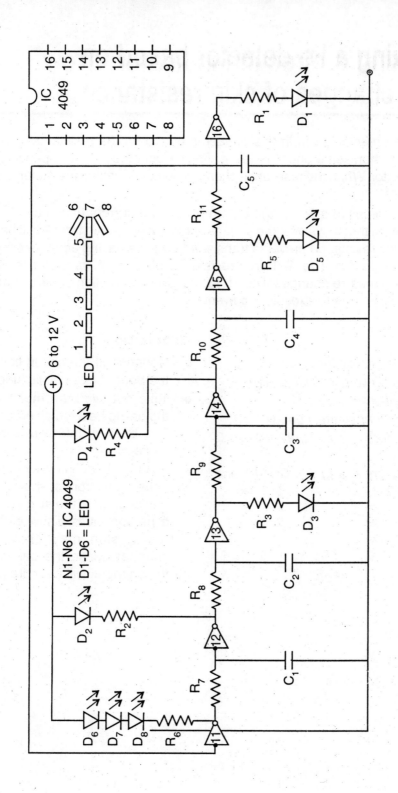

74. Making a lie detector based on the changes of skin resistance

The lie detector given in this project is based on the resistance changes caused when some one tells lies. It simply illustrates the basic principle of a lie detector. When the resistance changes due to sweating, the meter reading changes.

The circuit is a two stage DC amplifier. The current that is amplified is the current that flows over the surface of the person's skin from 9 volt battery. The 1 kΩ resistance prevents transistor leakages having adverse effect on circuit operation. The 470 ohm resistance also stops transistor leakages and temperature changes. The 10 kΩ provides protection against burn out of the meter due to excessive current.

You Require

- *One PCB or plastic lunch box*
- *One voltmeter with 2.5 volt scale*
- *A 9 volt battery*
- *One pot (potentiometer) 50 kΩ*
- *Four Resistors – 2.1 kΩ, 47 kΩ, 1 kΩ and 470Ω*
- *Two transistors T_1 and T_2 – BC149, one diode IN 4001*

What To Do

- Connect all the circuit components on the bottom of a plastic lunch box.
- Take out two bare metallic wires from the box to be used as probe for the skin. The back of the skin is a convenient place.
- Ask the questions to a person to test his honesty. An honest answer will not bring any change in the skin resistance but if a person gives dishonest answer and fears, skin may show resistance changes. Make the person emotional, the changes will be more pronounced.

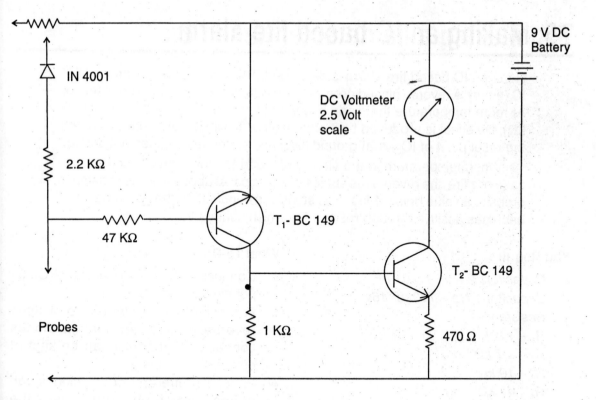

Components

1. 50 kΩ pot
2. Diode IN 4001
3. Resistances
   2.2 kΩ
   47 kΩ
   1 kΩ
   470 Ω
4. Probes (Red and Black)
5. Transistors T_1 and T_2 – BC – 149
6. Voltmeter – 2.5 volt scale
7. Battery – 9 vDC

75. Making an IC based fire alarm

Here, an IC based fire alarm is described. This circuit is made around 555 IC (timer). A reverse biased germanium diode is used here as a heat sensor. At room temperature the diode has very high reverse resistance (over 10-$k\Omega$). No effect is produced by it on transistor T_1 which conducts and beeps the reset pin 4 of IC – 1 at ground level and so the alarm does not sound.

When temperature in the vicinity of diode D_1 (the sensor) increases in case of a fire, the reverse resistance of D_1 drops. At about 70°C its resistance drops to a value below 1 kΩ. This stops T_1's conduction, the IC's reset pin 4 becomes positive through resistor R_2, which sounds the alarm.

You Require

- Capacitors 2 Nos – 0.01 μF
- Capacitor 1 No – 100 μF/16v
- Resistances
  R_1 – 1 kΩ
  R_2 – 4.7 kΩ
  R_3 – 10 kΩ
  R_4 – 47 kΩ
- VR_1 – 100 kΩ (preset)
- IC – 1 – IC 555 8 pin IC base
- Transistor T_1 – BC 548
- LS – 8Ω (speaker)
- Diode D_1 – DR 25 Germanium diode.
- PCB

What To Do

- Connect all the components on PCB as per circuit diagram. .
- For installation of alarm, two or three reverse biased germanium diodes connected in parallel can be kept at different locations.
- In case of fire any of the diodes can sense the heat and raise the temperature.

Note: DR25 diode works as a sensor but base emitter junctions of such germanium transistors as AC 128, AC 188 or 2 N360 can also be used.

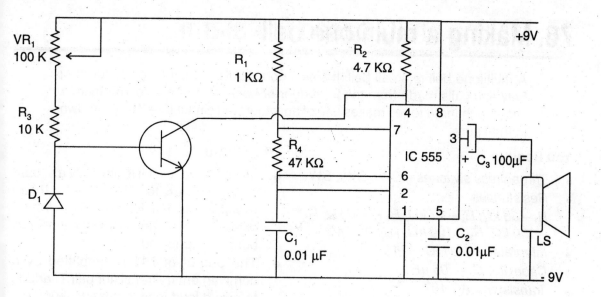

Components

1. Transistor T_1 – BC 548 – 1
2. Capacitors C_1 & C_2 – 0.01 µF – 2
3. Capacitor C_3 – 100 µF/16v – 1
4. Resistances
   R_1 – 1 kΩ
   R_2 – 4.7 kΩ
   R_3 – 10 kΩ
   R_4 – 47 kΩ
   VR1 – 100 kΩ (preset)
5. IC 555
6. Loudspeaker 8 Ω
7. Diode D_1 DR 25 Ge
8. PCB

76. Making a multitone bell

A multitone bell is a call bell. It saves up and down trips by distinguishing between calls made from the front door or back door or a side entrance. A circuit with three tones makes use of an integrated circuit IC – 741 as shown in the figure.

You Require

- Three press switches – SW_1, SW_2, SW_3
- Resistances
  R_1 – 33 kΩ, R_2 – 47 kΩ, R_3 – 100 kΩ, R_4 – 100 kΩ, R_5 – 100 kΩ, R_6 – 100 kΩ
- Integrated circuit IC – 741
- Capacitor C_1 – .05 μF
- Transistor – AC 187
- Transistor – AC 188
- Capacitor C_2 – 100 μF
- Battery 9vDC
- Loudspeaker – 8Ω
- A small lunch box of plastic or PCB

What To Do

- Multitone bell circuit is basically a square wave generator in which three resistances R_1, R_2 and R_3 produce different tones when any of the three call buttons is pressed.
- The output of 741 is amplified by a complementary transistor pair T_1 and T_2 to give a loud tone in the speaker.
- Make all connections as per the circuit diagram in the plastic lunch box or PCB.

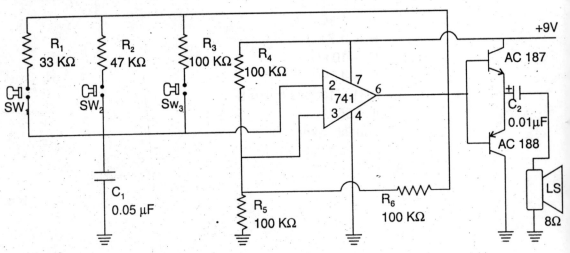

Components

1. Switches SW_1, SW_2, SW_3
2. Resistances
   R_1 – 33 kΩ
   R_2 – 47 kΩ
   R_3 – 100 kΩ
   R_4 – 100 kΩ
   R_5 – 100 kΩ
   R_6 – 100 kΩ

3. Capacitors C_1 – 0.05 μF and C_2 – 100 μF
4. Transistor AC – 187, AC – 188
5. IC – 741
6. Loudspeaker – 8 Ω
7. Battery – 9vDC

77. Making a water-level indicating alarm

When you fill your water tank by running a motor, you do not know when it is filled. Thus water level alarm becomes a basic requirement for every home. When we start a motor to fill up the tank, this alarm gives the indication of sound. When the tank is filled with water, alarm is on. When you use water and the two wires do not make a contact with water, the alarm goes off.

You Require

- Resistance R_1 – 100 kΩ
- Resistance R_2 – 56 kΩ
- Resistance R_3 – 10 kΩ
- Transistor T_1 – BC 548
- Transistor T_2 – BC 558
- Capacitor C_1 – 0.01 μF
- Battery 3 vDC (Two pencil cells)
- Loudspeaker – 2.5 inch, 8 ohm
- Cell connector
- PCB
- Two wires to make a contact with water

What To Do

- In this project we use only 3 volt pencil cells to conduct electricity. Very low voltage is applied to water to act as a conductor.
- We make use of two transistors, one is NPN and other is PNP. Both these are connected as a complementary oscillator.
- Resistance R_1 gives a positive low voltage to base of NPN.
- Resistance R_1 and C_1 are connected as a feedback circuit. Feedback circuit produces an oscillation and alarm gives a sound.

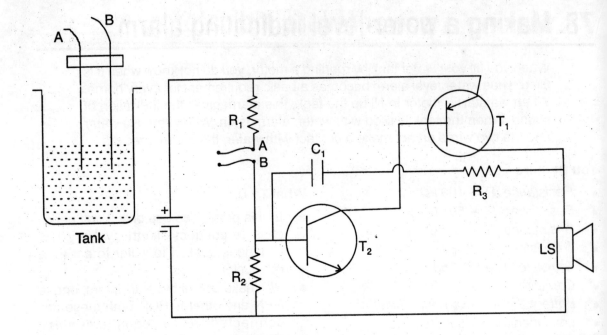

Tank

Components

1. Resistances
   R_1 – 100 kΩ
   R_2 – 56 kΩ
   R_3 – 10 kΩ
2. Capacitor C_1 – 0.05 µF
3. Transistor T_1 – BC 558
4. Transistor T_2 – BC 548
5. Speaker 2.5″, 8Ω .
6. 3 vDC (Two pencil cells)
7. Cell connector
8. Water level wires
9. PCB

78. Making a power failure indicator

Supply failures are very normal these days. This circuit indicates power failure by glowing LED. Some of the digital systems need a continuous power supply to ensure correct operation. A digital clock is an example which needs a continuous supply. If the power fails, we have to reset the clock for correct time.

You Require

- Reset switch
- Diode D_1 – IN 4001
- Capacitor – 0.1 μF
- Resistances R_1 – 100 kΩ, R_2 – 10 kΩ
  R_3 – 10 kΩ, R_4 – 1 kΩ
- IC – 741
- LED – Red
- Battery 0.5 to 12 v
- PCB

Components

1. Reset switch
2. Capacitor – 0.1 μF
3. Resistances R_1 – 100 kΩ
   R_2 and R_3 – 10 kΩ, R_4 – 1 kΩ
4. IC – 741
5. Diode D_1 – IN 4001
6. LED
7. Battery 0.5 to 12v (adaptor)

What To Do

- Make connections as per the circuit diagram on the PCB.
- In case of power failure the LED will glow.
- When the supply is switched on, the voltage at pin 2 of 741 is 0.6 volts lower than the supply voltage. Pressing the reset button makes pin 3 voltage higher than that at pin 2 and the output swings high. Positive feedback via R_2 makes the circuit latch in this state. The LED is therefore not lit.
- When the supply is interrupted, all voltages fall to zero. Upon restoration of the supply, the inverting input is immediately pulled up to previous voltage by diode D_1. However, capacitor C_1 being uncharged, holds the voltage at the non-inverting input low. The circuit stays in this state still till reset button is pressed again.

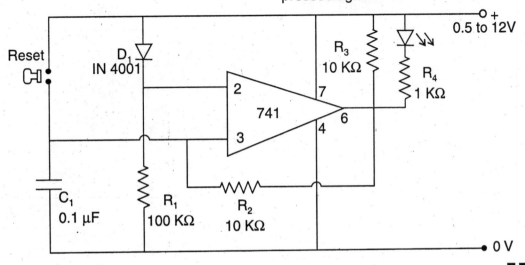

79. Making dancing or disco lights

This type of dancing or disco lights are commonly seen in the movies or clubs. These lights have charming colours dancing on the shirts or coats of performers. You can make the dancing light with the help of this project. This is an inexpensive project and easy to make.

You Require

- *Transistors – 2 Nos – BC 148*
- *Capacitors – 2 Nos – 10 µF/12v*
- *Resistances – 4 Nos – 47 kΩ*
- *Battery – 9vDC*
- *LED – 6 Nos – 3 Red and 3 green*
- *PCB*

Components

1. Transistors – 2 nos – BC 148
2. Capacitors – 2 nos – 10 µF/12
3. Resistances – 4 nos – 47 kΩ
4. LED – 6 nos – 3 green, 3 red
5. Battery 9vDC
6. PCB

What To Do

- Connect all the components on PCB as shown in the circuit diagram.
- The positive wire of first LED in each series should be connected to the positive terminal of the battery.
- Try to make it on a small PCB so that it can be pinned with the pocket of the shirt.
- After assembling all the components as per the circuit, as soon as you connect the battery, you will see the charming green and red colours. You can put a switch in the battery circuit.

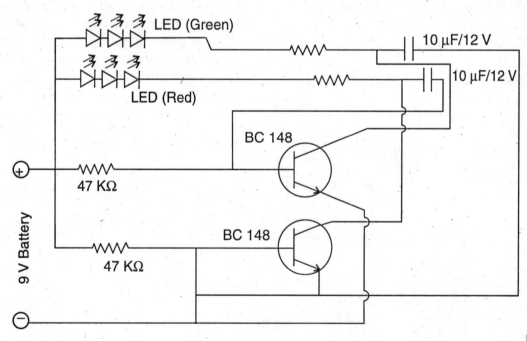

80. Making a battery operated tube light

This project is very useful for cities like Delhi where power failures are quite frequent. At the time of power failure your circuit will work and tube light will be on. One can study in tube light during examination da

You Require

- *Transistors – AC 127 NPN*
- *D.C. source – 4.5 volt (battery)*
- *Potentiometer – 5 kΩ*
- *Tube light – long 1.06 m (40 watts)*
- *Step down transformer*
  Primary 220 to 240 volts
  Secondary 6 volts

What To Do

- Make the connection of components as shown in the circuit diagram.
- We start from the negative terminal of the source. We connect this terminal of the emitter of the transistor AC 127.
- The positive terminal of the battery is connected to the middle of the secondary circuit of the transformer.
- The base of the transistor is connected to any terminal of the potentiometer. The other terminal of the pot is connected to the last terminal of the secondary of the transformer.

- The first terminal of the secondary of the transformer is connected to the collector of the transistor.
- Now the two ends of the primary of the transformer are connected to the two ends of the tube light.
- A heat sink should be used with AC 127 for proper heat dissipation otherwise transistor may get damaged. For more powerful light we can use transistors of value OC 26, OC 27, AD 139, AD 149 etc.

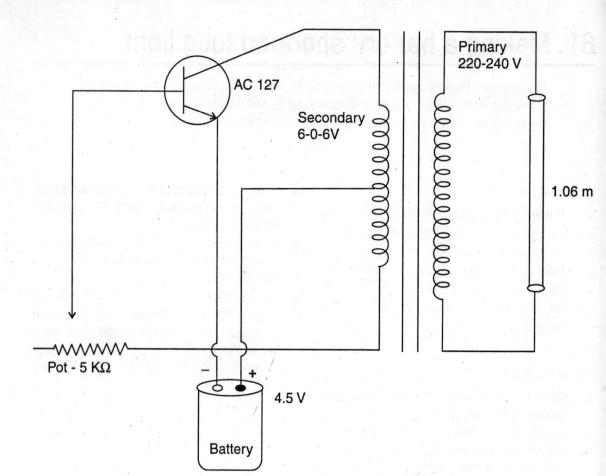

AC 127

Secondary
6-0-6V

Primary
220-240 V

1.06 m

Pot - 5 KΩ

− +

4.5 V

Battery

Components

1. Transistor – AC 127
2. Potentiometer – 5 kΩ
3. Battery – 4.5 v
4. Tube light
5. Set down Transformer
   Primary 220 – 240 volts
   Secondary 6 – 0 – 6 v
6. PCB or a box

81. Making a two transistor radio

The circuit of a two transistor radio shown here makes use of a few components but is complex in operation. Nowadays transistors are cheap but there was a time when these were costly. With this circuit you can make a cheap transistor radio.

You Require

- *Antena coil medium wave – L_1*
- *Diodes D_1 and D_2 – Two – IN 4001*
- *Transistor NPN, T_1 and T_2 – 194 B and 148 B*
- *VC – 1 2x Gang*
- *Capacitors C_1 and C_2 – 0.01 µF*
- *Capacitor C_3 – 470 pF*
- *Capacitor C_4 – 1 µF*
- *Resistances – 3 nos*
  R_1 – 560 Ω
  R_2 – 47 kΩ
  R_3 – 15 kΩ
- *Capacitor C_5 – 0.02 µF*
- *Inductance coil*
- *Ear phone – 8 Ω*
- *PCB*

What To Do

- Make the connections on PCB as shown in the circuit diagram.
- Wound 150/200 turns with a thin wire (slightly thicker than hair) enamelled copper wire on the former pasted/premounted on PCB.
- The coil L_1 is made from 88 turns of enamelled copper wire of 32 swg, on a 5/18 inch ferrite rod about 4 inch long with a tap of 8 turns.
- The tuned circuit is made up from VC_1, 80 turns of L_1 and C_1.
- The 8 turns act as an auto transformer giving a good match to the base of T_1. The transistor amplifies the RF which is fed to T_2 acting as an emitter follower.
- A transistor BC 108 will also work in the circuit.
- All the transistor tuning type capacitors will work as VC_1.

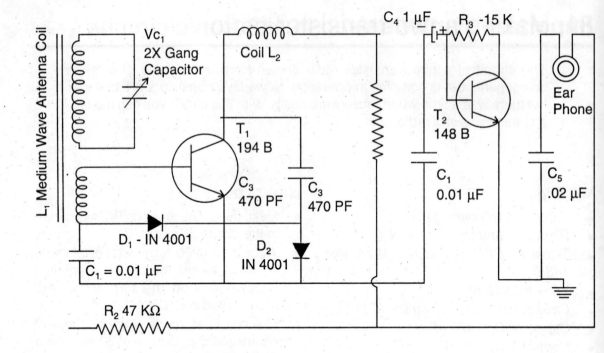

Components

1. Medium wave autenna L_1
2. Two diodes D_1 & D_2 – IN 4001
3. T_1 and T_2 transistors 194 B and 148 B
4. VC1 2x gang
5. Capacitor C_1 and C_2 – 0.01 µF
6. Capacitor C_3 and C_4 – 470 pF and 1 µF
7. Resistances R_1 – 560 Ω, R_2 – 47 kΩ,
   R_3 – 15 kΩ
8. Capacitor C_5 – 0.02 µF
9. L_2 – Inductonce Coil
10. Ear Phone 8 Ω

Important symbols used in electronic circuits

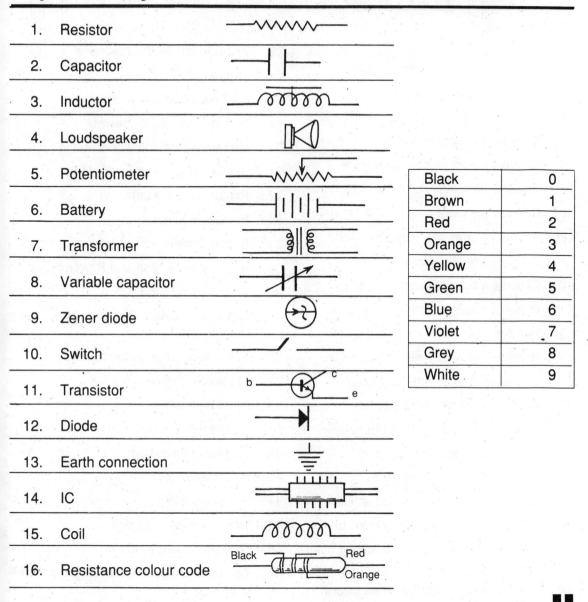

1.	Resistor	
2.	Capacitor	
3.	Inductor	
4.	Loudspeaker	
5.	Potentiometer	
6.	Battery	
7.	Transformer	
8.	Variable capacitor	
9.	Zener diode	
10.	Switch	
11.	Transistor	b, c, e
12.	Diode	
13.	Earth connection	
14.	IC	
15.	Coil	
16.	Resistance colour code	Black, Red, Orange

Black	0
Brown	1
Red	2
Orange	3
Yellow	4
Green	5
Blue	6
Violet	7
Grey	8
White	9

Note: For any item related to any project please contact the following firm.

OPTOCHEM INTERNATIONAL
19/15/1, West Moti Bagh
Opposite Sarai Rohilla,
Police Station, Delhi-35

Additional projects you can try

1. To make a battery from the juices of citrus fruits such as lime, orange, etc.
2. To make a solar cooker
3. To make a hearing aid
4. To make a telephone set
5. To make a compound microscope
6. To demonstrate optical communications
7. To study the plant adaptations for water
8. To make a code practice oscillator
9. To make a fancy television
10. To make a table cigarette-lighter
11. To make a toy transmitter
12. To determine the calorific value of fuels
13. To make logic NDR with LED display
14. To make logic DR with LED display
15. To make AND with LED display
16. To make logic NAND with LED display
17. To make an electronic fish caller
18. To make LED strobe lights
19. To make an electronic woodpecker
20. The effect of magnetism on growth
21. The effect of ultrasonic waves on plant growth
22. The effect of UV light on plant growth
23. Paper chromatography
24. Desalting water by magnetic and static fields
25. To extract caffiene from tea leaves
26. To demonstrate fluorescence
27. To make a water timer
28. To make a dry cell charger
29. To make a mosquito repellant
30. To make a classroom addressing system
31. To convert tea water into black ink
32. To make a fire fountain
33. To make nylon fibres
34. To make photographic paper
35. To make Ram, Sita and Rawana toys
36. To make a Vasudeva cup
37. To make a wheel revolving by thermal expansion
38. To make an aneroid berometer
39. To make a mini-electric fan
40. To make a microphone
41. To make an electric heater
42. Silver electroplating
43. Nickel electroplating
44. Chrome-plating
45. To make a dancing doll
46. To make a burglar alarm
47. To make a liquid temperature control device
48. To demonstrate electrowriting
49. To make a sound battery
50. To make a police siren
51. To make a solar slide projector

Electronics Projects for Beginners

—A.K. Maini

Using Easy Available Electronic Components

Electronics Projects For Beginners has been written for a wide cross-section of readers in ITI and Diploma level students looking for suitable project exercise to be done as a part of their academic curriculum, hobbyists and electronics enthusiasts. The book has been written in a very simple language with large number of illustrations and contains in one volume both the detailed account of projects as well as the relevant theory in the form of two introductory chapters in order to enable the users of the book fully understand what they are attempting to build. The book contains 40 projects in all complete with comprehensive functional description, Parts list, Construction details such as PCB and Components' layouts, Testing guidelines, suitable alternatives in case of uncommon components and lead/pin identification guidelines in case of Semiconductor Devices and Integrated Circuits (ICs). The first two introductory chapters contain a lot of practical information. The first chapter gives application relevant information in case of electronic components such as Resistors, Capacitors, Coils, Transformers, Diodes, Transistors, LEDs, Displays, SCRs, Opamps, Timers, Voltage Regulators and General purpose digital ICs such as Gates, Flip flops, Counters etc. The second chapter on constructional guidelines gives a brief account of Soldering techniques, PCB making guidelines and guidelines to using general purpose test and measuring equipment such as Ammeters, Voltmeters, Analog and Digital multimeters, Oscilloscopes etc. The two introductory chapters followed by detailed description of 40 tested projects thus make the book a true self-learning guide for those who wish to construct electronics projects.

Big Size • Pages: 196
Price: Rs. 96/- • Postage: Rs. 20/-

FUN, FACTS, HUMOUR, MAGIC, MYSTERY & HOBBIES

Strange But True Facts
80/-
Demy size, pp: 184

FUN WITH NUMBERS
40/-
Demy size, pp: 115
also available in Hindi

101 BRAIN TEASERS
48/-
Demy size, pp: 152
also available in Hindi

Incredible but True
36/-
Demy size, pp: 112
also available in Hindi

501 FASCINATING FACTS
40/-
Demy size, pp: 104
also available in Hindi, Bangla, Kannada & Assamese

501 ASTONISHING FACTS
36/-
Demy size, pp: 115
also available in Hindi

How to solve Crossword Puzzles
60/-
Demy size, pp: 104

Amusing Anecdotes on Indian Red Tape
80/-
Demy size, pp: 176

Rib-Tickling JOKES
48/-
Pages: 128

THE WORLD'S BEST PROFESSIONAL JOKES
60/-
Pages 120

MEDICAL JOKES & HUMOUR
48/-
Pages: 152

ARMOUR OF HUMOUR
40/-
Demy size, pp: 128

DEFT DEFINITIONS
48/-
Demy size, pp:

Amusing Encounters of Daily Life
68/-
Demy size, pp: 124

The Funniest Tales of Mullah Nasruddin
48/-
Pages: 144
(Also in Hindi)

50 WITTIEST TALES OF BIRBAL
48/-
Pages: 120

UNWRITTEN FLAWS OF INDIAN BUREAUCRACY
295/-
Demy size, pp: 248
(Hardbound)

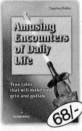

New
The Red Monster
60/-
Demy size, pp: 104

New
The Witches of Waitiki
68/-
Demy size, pp: 176

New
The Woman in White
80/
Demy size, pp: 1

101 Magic Tricks
Easy to learn & perform
88/-
Big size, pp: 112
(Full colour book)

MAGIC FOR FUN
40/-
Demy size, pp: 112
also available in Hindi, Kannada and Marathi

MAGIC for CHILDREN
48/-
Demy size, pp: 124
also available in Hindi

MAGIC for YOU
40/-
Demy size, pp: 124
also available in Hindi,

HOW TO DRAW CARTOONS
60/-
Demy size, pp: 124
also available in Hindi

Drawing and Painting Course
60/-
Big size, pp: 120
also available in Hindi

POSTAGE: RS. 15 TO 25/- EACH

SELF-IMPROVEMENT

How to Remain Ever Happy
Tips to relieve yourself from Stress, Tension and Anxiety

68/-

Demy size, pp: 160
Also available in Hindi and Bangla

How to Control MIND and be Stress-Free

60/-

Demy size, pp: 136
Also available in Hindi

HOW TO CONTROL ANGER — THE DEADLY ENEMY

60/-

Demy size, pp: 64
Also available in Hindi

HOW TO OVERCOME FEAR

48/-

Demy size, pp: 80
Also available in Hindi

Correct Etiquette & Manners for all occasions

68/-

Demy size, pp: 156

Self ANALYSIS
The Practical workbook based on DIANETICS techniques

88/-

Demy size, pp: 290

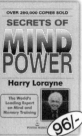

SECRETS OF MIND POWER
Harry Lorayne
The World's Leading Expert on Mind and Memory Training

96/-

Demy size, pp: 184

Freedom from Self-punishing thoughts

96/-

Demy size, pp: 192

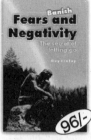

Banish Fears and Negativity
The secret of letting go

96/-

Demy size, pp: 240

Be Confident & Fearless
How to overcome fears and empower yourself

120/-

Demy size, pp: 240

The Portrait of a Complete Woman
A guide to woman's personality development

120/-

Demy size, pp: 304

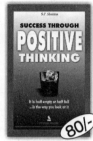

SUCCESS THROUGH POSITIVE THINKING
It is half empty or half full ...is the way you look at it

80/-

Demy size, pp: 180

The Meaning to KNOW THYSELF
How to Attain Happiness and Good Life

60/-

Demy size, pp: 128

Off-loading Stress At Work Place
Relaxation techniques through simple exercises, postures breathing & meditation at work place

80/-

Demy size, pp: 218

PEACE of MIND

80/-

Demy size, pp: 174

JOURNEY INTO A FULFILLING LIFE

68/-

Demy size, pp: 176

From Despair to Joy
Tips to turn negative thinking into positive

80/-

Demy size, pp: 140

SMILES ...miles away from stress
Self-help manual identifying stressors in daily life and how to learn to cope with them

60/-

Demy size, pp: 140

Build Self-confidence
Practical guidelines to personal and professional success

80/-

Demy size, pp: 155

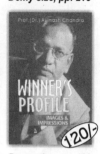

WINNER'S PROFILE
IMAGES & IMPRESSIONS

120/-

Demy size, pp: 128

Understanding Emotional IQ

68/-

Demy size, pp: 176

What's your Emotional IQ

68/-

Demy size, pp: 176

BE A WINNER
How to come out a winner in the face of heavy odds

80/-

Demy size, pp: 136

UNLEASH YOUR HIDDEN POTENTIAL
Over 40 self-analysis modules to help you achieve excellence in your career.

80/-

Demy size, pp: 176

POSTAGE: RS. 15 TO 25/- EACH

How to Develop The Right Attitude

New

68/-

Demy size, pp: 96

J. Krishnamurti Demystified

New

80/-

Demy size, pp: 176

Harry Lorayne
How To Develop A SUPER POWER MEMORY

New

120/-

Big size, pp: 168

SELF-IMPROVEMENT

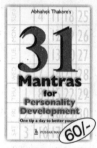
Demy size, pp: 104

Demy size, pp: 176

Demy size, pp: 112

Demy size, pp: 136

Demy size, pp: 176

Demy size, pp: 112

Demy size, pp: 176

Demy size, pp: 128

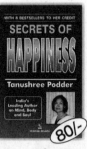
Demy size, pp: 192

Demy size, pp: 144

Demy size, pp: 376

Demy size, pp: 168

Demy size, pp: 160

Demy size, pp: 160

Big size, pp: 160
(In 2 colour)

Demy size, pp: 160
(In 2 colour)

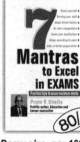
Demy size, pp: 160

Demy size, pp: 14_

Demy size, pp: 192

Demy size, pp: 176

Demy size, pp: 96

Demy size, pp: 144

Demy size, pp: 128

Big size, pp: 128

Big size, pp: 312

POSTAGE: RS. 15 TO 25/- EACH

CAREER/STUDENT DEVELOPMENT/MANAGEMENT

Become a Successful SPEAKER — Don Aslett
68/-

Demy size, pp: 136

SUCCESS SECRETS — A COMMON-SENSE GUIDE TO LIFELONG ACHIEVEMENT — MERRILL DOUGLASS
120/-

Demy size, pp: 256

Youngsters' Guide for PERSONAL DEVELOPMENT
68/-

Demy size, pp: 120

too You can Become Rich — Dynamic self-action plan to take-off.
80/-

Demy size, pp: 128

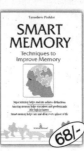

SMART MEMORY — Techniques to Improve Memory
68/-

Demy size, pp: 138

How to Motivate Others to turn them into super performers
80/-

Big size, pp: 128

SOLVE YOUR PROBLEMS ...The Birbal Way
68/-

Demy size, pp: 200

A guide for youngsters TEENS TO TWENTIES
68/-

Demy size, pp: 120

Create your own SUCCESS STORY & Live Life King-Size
80/-

Demy size, pp: 120

Making Friends and doing business in Europe
96/-

Demy size, pp: 288

The Street Smart Salesman ...Making Opportunities Happen
88/-

Demy size, pp: 208

DON ASLETT How to be a favourite with your BOSS
80/-

Big size, pp: 106

20 Keys for SUCCESS in JOB & CAREER
80/-

Demy size, pp: 144

GREAT SPEAKERS AREN'T BORN — The Complete Guide to Winning Presentations — GEORGE KOPS & RICHARD WORTH
88/-

Demy size, pp: 200

Skills for Excellence — LUIS S. R. VAS
88/-

Demy size, pp: 184

How to be the Complete Professional Salesperson — Robert L. Shook
120/-

Demy size, pp: 248

James M. Bleech & Dr. David *New* Know-it-all Guide LET'S GET RESULTS NOT EXCUSES! — A LEADER'S GUIDE TO EFFECTING CHANGE IN THE CORPORATE WORLD
195/-

Big size, pp: 240

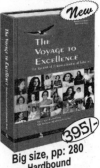

New THE VOYAGE TO EXCELLENCE
395/-

Big size, pp: 280 Hardbound

U.S. VISA MADE EASY — a practical guide
220/-

Big size, pp: 188

MARKETING WITH SPEECHES AND SEMINARS — Your Key to More Clients and Referrals
80/-

Demy size, pp: 176

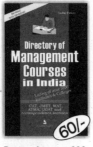

Directory of Management Courses in India — CAT, JMET, MAT, ATMA, UGAT and correspondence courses
60/-

Demy size, pp: 392

BEGINNERS' GUIDE TO JOURNALISM — Acquire better understanding of Communication Process
80/-

Demy size, pp: 128

Study & Immigration in U.S.A
95/-

Demy size, pp: 128

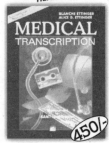

BLANCHE ETTINGER ALICE G. ETTINGER MEDICAL TRANSCRIPTION
450/-

Big size, pp: 472

POSTAGE: RS. 15 TO 25/- EACH

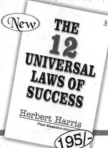

New THE 12 UNIVERSAL LAWS OF SUCCESS — Herbert Harris
195/-

Demy size, pp: 192

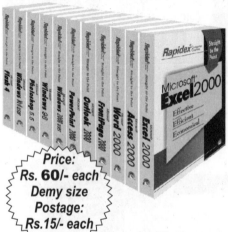
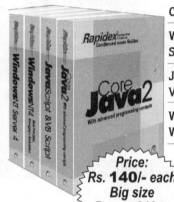

DICTIONARIES & ENCYCLOPEDIAS

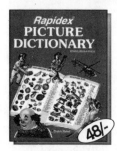

Big size, pp: 48

(In colour)

48/-

Demy size, pp: 136

60/-

Big size, pp: 231

120/-

Big size, pp: 58

(In colour)

72/-

Big size, pp: 98

(Double colour)

72/-

Big size, pp: 520

380/-

Big size, pp: 384

360/-

Demy size, pp: 344

120/-

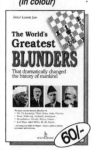

Demy size, pp: 128

60/-

Big size • pp: 52 (In 4 colour)
Deluxe Binding
Also available in Hindi

100/-

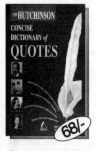

Demy size, pp: 352

68/-

Demy size, pp: 184

50/-

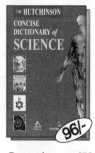

Demy size, pp: 456

96/-

Demy size, pp: 128

60/-

Demy size, pp: 152

24/-

Demy size, pp: 196

60/-

Demy size, pp: 128

60/-

Demy size, pp: 232

68/-

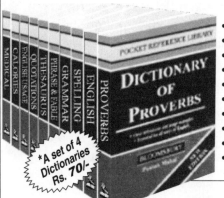

Bloomsbury Dictionaries

* Dictionary of Phrase & Fable
* English Thesaurus
* Spelling Dictionary
* Dictionary of English Usage
* Medical Dictionary
* Dictionary of Calories
* English Dictionary*
* Dictionary of Grammar*
* Dictionary of Proverbs*
* Dictionary of Quotations*

*A set of 4 Dictionaries Rs. 70/-

Pocket size • Pages: 256
Price: Rs. 30/- each • Postage: Rs. 10/- each

GENERAL KNOWLEDGE/SCIENCE

Children's Knowledge Bank
(in 6 independent volumes)

Your Prodigy Child is growing up. He is amazed and puzzled by the wonderful world around him. So many How's and Why's come in his mind. He expects rational and logical explanations to his queries. He is also not content to learn things by halves. And what happens if he does not get answers at all to his queries? He may start accepting anything and everything forced on to him. The making of a dull child begins thus.

Something has to be done right now, to nourish this budding scientist and

help him realize his full potential.

GET HIM 'CHILDREN'S KNOWLEDGE BANK' TODAY.

The series gradually builds up his foundation and IQ and acts as a tonic for his brain.

SAVE **20%**
Buy full set of 6 Vol. for Rs. 300/- in a gift box instead of Rs. 360/-

Published in 8 more languages: **Hindi, Bangla, Kannada, Tamil, Telugu, Gujarati, Malayalam, Marathi***

Big size • Price: Rs. 60/- each (Full set: Rs. 300/- instead of 360/-)
**Price: Rs. 68/- each ■ (Full Set: Rs. 340/- instead of Rs. 408/-)*

Children's Science Library
A set of 17 books

A unique library that helps your child build a scientific temper and logical approach. Once a child's fundamentals are clear, he is sure to excel in studies. It will supplement the basic school educaton and help the child develop a keen interest in Science and expand his horizons.

Books of the Series:

- The Earth • Life on Earth • The Universe • Energy
- Light • Sound • Transport • Plant Kingdom
- Communications • Animal Kingdom • The Human Body
- Minerals & Metals • Force and Movement
- Scientists & Inventions • Electricity & Magnetism
- Environment & Pollution • Development of Chemistry

Books of the Series:
- *The Earth • Life on Earth • The Universe • Energy*
- *Light • Transport • Plant Kingdom • Communications*
- *Sound • Animal Kingdom • The Human Body • Minerals & Metals • Force and Movement • Scientists & Inventions*
- *Electricity & Magnetism • Environment & Pollution*
- *Development of Chemistry*

you save **20%**
Pay Rs. **350/-** instead of Rs. 425/- for complete set of 17 books priced Rs. 25/- each

Big size, Price: Rs. 25/- each
Postage: Rs. 15/- each
Postage: Free on 6 or more books
Also available in Hindi